On Freedom:

Organizational Science Examined Philosophically

By

Peter Gibson Friesen

PM Library

copyright © 2016 by Peter Gibson Friesen

ISBN: 978-0-9860600-9-0

All rights reserved. No part of this book may be used or reproduced in any manner whatsoever without written permission, except in the case of quotes for personal use and brief quotations embedded in critical articles or reviews.

PM Library
an imprint of Poetic Matrix Press
John Peterson, Publisher
www.poeticmatrix.com

Acknowledgments

I would like to gratefully acknowledge the following individuals for showing interest in this work, and allowing it to become part of their thinking about social science and what we refer to as "organization."

Alison Brause, David Boals, Maurice Bisheff, and Akash Mittal rate highly among the students (and teachers) of the organizational science we refer to as Requisite Organization. They have encouraged me to persist in the clarification of these ideas despite the fact that there really isn't a well-defined audience, even a niche audience. The niche seems to devour its surroundings.

I am grateful, of course, to Elliott Jaques for communicating with me about this work in its early stages—way back in the early 1980s. His recognition of information processing strategies as a viable descriptive matrix, along with his excitement about them, were very motivating. What a teacher he was. I am also grateful to his surviving spouse Kathryn Cason, for reading and acknowledging the value of some of the ideas presented here in their more primitive form.

I am deeply grateful, perhaps beyond words, for my father Ernest Friesen's support in the completion of this book. He is responsible for really great strides in the development of a science of organization and management, and its application to the administration of justice. If it were not for his engagement with me in the discussion of abstract ideas that we both own together, including the human as a moral and spiritual presence in the universe, this book would not have been written.

Last, I am grateful to John Peterson, who has been following various draft manuscripts engaging the ideas here since the mid-1990s. He is a poet and philosopher who examines life compassionately and owns the company that is now publishing this book. His thoughts and encouragement have had a positive impact on the content of it.

Contents

On Freedom:

Organizational Science Examined Philosophically

Acknowledgments iii

Contents v

Preface viii

1. Scientific Freedom 2

2. Intellectual Freedom 36

3. Moral Freedom 74

4. Political Freedom 112

5. Spiritual Freedom 152

Postscript 188

Appendix: The system of notation that marks
the contours of a Logic of Maturation 194

Endnotes 214

Author Biography

On Freedom:

Organizational Science Examined Philosophically

Preface

Peter Gibson Friesen

There is something pleasant in the realization that upon the completion of writing an abstract and contemplative treatise I am tasked to write a "preface" so that others might track with the development of its argument. It thus extends as a gesture of good will from a perspective which was once reflective and insular, now changed by the prospect that others could trouble themselves to understand what these five essays on freedom mean.

Not that they necessarily "mean" anything, or that I might change what they are through insertion. It might help, nonetheless, for me to share what motivated the devotion of many years of labor to this task.

This work did not start as a treatise on freedom, but rather, as a fascination with discoveries of "fact" about human institutions. These discoveries occurred within a movement of research occurring over the latter half of the twentieth century. It was once called "stratified systems theory" by its progenitor Elliott Jaques (deceased 2003), because of its focus on "hierarchy" in management structures. It later became "requisite organization" as the research matured into a systematic comment on the "health" of bureaucratic structures. The name "requisite" signified a move from organization as something imposed on human nature to something that agrees with it.

The "facts" unearthed within this movement were remarkable, revealing consistancy in the way human collaborative endeavors organize hierarchically—across all contexts of law, economic production, institutional output and culture. These facts spawned a generation of theory, not just about what human organization is, but about what human beings are. We are, it seems, engaged in stages of growth, and we experience fairly predictable changes in the way we solve problems. These changes represent progressive improvement of control over one's work environment, and thus do much to account for the development of hierarchy in human institutions. I have given some thought to the explanation of how and why this happens, and have written it down.

However, an explanation of the development of hierarchy hardly explains why this book would have as its title <u>On Freedom</u>. There are other ingredients in the formation of this treatise worth mentioning.

The examination of "organization" and "hierarchy" through a philosophical lens moves away from the assignment of specialized roles within a given work system, and undertakes a more abstracted evaluation of mind in a state of organization. To some degree, organization is something imposed by mind on its surroundings, but this imposition standing alone is restrictive. Organization changes, and not so much because it is programmed to do so, but because we want it to. This gives one much to think about both epistemologically and ethically. Organization is therefore philosophically provocative, and a worthy subject of philosophical inquiry.

In the process of making this inquiry, "freedom" begins to stand out as interesting in its own right, and benefits from theories about how individuals develop the capacity to think, feel, cooperate and worship. The fact that we experience and break through confinements, and become better than what we were before, is a more interesting way to examine human development than drawing status comparisons between persons. An emphasis on the transcendence of boundaries—not their obduracy—brings the examination of human development close to what most people mean when they use the term "freedom." "Freedom" denotes ethical elevation along with the easing of restriction.

No wonder that talk of "freedom" is woven into our resistance to immoral tyrants, and into narratives of spiritual transformation. We might get carried off by emotion, but science encourages us to view the human being with circumspection. Indeed, scientific skepticism is famous for debunking inspirational tales of transcendence and healing, told and retold because we like to think that existence is something more than competition among rivals. We are therefore compelled to ask if science is somehow responsible for a debased view of humanity, and whether it is incompatible with freedom.

This question launches my first essay on freedom, entitled Scientific Freedom. There I contend that science does not debase our view of humanity, but to the contrary, prevents that view from becoming congested with superficial views, and moves forward with the belief that we were meant to know the truth. Scientific skepticism, an attitude which demands patience and accountability in one's theoretical orientation, is far from depressed. It is energized and excited by facts, unexplained facts which lead us to doubt our theories.

Facts prompt a reevaluation of human consciousness, and lead us toward a behavioral science receptive to something often referred to as "spiritual." This is particularly the case if there are—in a manner of speaking—"facts" which tend to support theories that elevate human existence from a material plane and reposition it within an ethical universe.

I am informed that the most difficult essay is Intellectual Freedom (Chapter 2). I struggled with it, perhaps more than necessary, and made many efforts to make it comprehensible to non-logicians. A little history might help.

Upon my first exposure to Jaques' work I endeavored to master theoretical essays then being written on a potential relationship between operations important to logical/mathematical systems, and the hierarchy of work roles in bureaucratic organizations. As I worked on these forms, it seemed that it might be productive to use information processing strategies related to data search—instead of symbolic logic—as explanatory media for work identified with various management strata. The experiment proved successful, as "information search" proved to be a simpler and more effective way to describe how intellect sequentially phases into the mastery of a given environment.

The phrase "information processing" places emphasis on the way particles of data are identified and organized by the human intellect. Typically, as complexes of data become more organized, one is better able to acquire what one seeks from that environment. Phases—qualitative shifts of one's strategic efficiency in the navigation of a given information system—occur because the mind acquires and integrates new strategic tools in the selective organization of data (information). This takes time, as does the acquisition of skills anywhere, and thus accounts for the time it takes for a career to progress upward through an institutional hierarchy.

What came as a bit of a surprise is that information processing strategies seemed—in addition to providing effective descriptors for management problem solving behavior—effective in the definition and clarification of concepts foundational to first order propositional logic (the logic studied in an undergraduate academic curriculum). It also functions effectively as an imperative logic.

There wasn't anything new about the use of an information processing metaphor to describe behavior. What was new was that we could use this metaphor within a calculus to effectively describe both cognitive development (learning) and organizational hierarchy. This might, therefore, be regarded as a shift toward a "science" of organization—where rules of motion applicable to learning are also rules of order applicable to the formation of institutions, all reduced to a fairly simple calculus.

In my second essay on "Intellectual Freedom," there is an opportunity to examine in some depth how the creation and management of information simply and naturally defines moments of breakthrough—events which correlate to terms used in the logic taught in academic classrooms. These occur through the sequential integration of a finite set of skills, each of which gives us more control over a complex environment. Freedom is not about taking something away, but of adding to what we take for granted. Growth and learning—and the freedom associated with them—are a result of affirmations.

There is more to it than that, of course, but it is important to acknowledge that "freedom" as a word involves the intervention of value or interest through which selection and choice occurs. Freedom is not so much about having a choice as it is about achieving and exercising choice through the affirmation of interest. The world without value is void, and interest makes the identification and selection of objects possible—objects as organized information. Freedom is thereby selecting a whole universe into existence.

I have been told more than once that the notational system I have used to express the logic of intellectual freedom (and information processing) is not familiar. It may therefore be useful to know that I appropriated the approach to symbolic representation from G. Spencer Brown, a mathematician who Bertrand Russell described as "genius." He developed a calculus based on the pictorial representation of a "distinction" drawn as a circle. Essential to his view of distinction, is the logical role of interest—i.e., distinction as motivated intervention. In order to assist in the reading of this essay, I have added an appendix to this book, explaining the notation and its debt to G. Spencer Brown.

Thinking of the intellect as a process that includes the assertion of human interest creates problems on other fronts. For several millennia, philosophers have identified human "reason" as the former of moral judgments because it is neutral, unmoved by personal attachment, and can thus apply moral principles with categorical certainty. But that is not consistent with the experience of morality. Even if we deduced actions from moral principles, we would still lack explanation of how we came to embrace such principles at all. I am much impressed in my life experience that we don't think them up, but come to feel them as attached to who we are. They come into being gradually through much iteration with other people.

Perhaps it is the realization that intellect is of its nature not ethically neutral that takes one past the intellect in order to define "moral freedom," the subject matter of my third essay. The words, "information processing," are inadequate to describe the activity of mind in handling other people. "Processing" suggests a mind which is merely reactive to data, and moral thinking is dominated by fictional scenarios, narratives rehearsed in storytelling, through which the mind practices the integration of socially harmonious compromise positions.

Moral freedom represents success in the appropriation of narratives unique to a given set of social challenges, of internalizing values that permit one to function gracefully within a society.

One thing the reader is apt to find interesting in the essay on Moral Freedom (Chapter 3), is that the plot forms of mythical narrative endorsed as universal within the field of literary criticism—Comedy, Romance, Tragedy and Irony—are descriptive of the four phases of information processing noted in the essay on intellectual freedom. The plots are there, however, to enable the experience of values essential to the definition of selfhood, and the "self"—as an object of study—is considerably more elusive than any complex system in an exterior world one cares to bring to the gaze of intellect.

This is a very important point in the understanding of moral judgment, and thus the analysis of moral freedom. The engagement of the intellect in the task of forming and knowing one's self is labor not much unlike that of a pack mule walking after a carrot that his rider dangles out in front of him, except that it does for many reach a point of satisfaction—which I think of as "moral" satisfaction. There are phases to moral development but they occur mythically in the parables and episodes designed to allow us to experience value, and do not track with the comprehension of our physical and social environment. The truth is—one may be quite advanced intellectually and yet quite infantile in his or her moral development.

To some extent, the moral journey is lonely, but there are significant social interests supporting the engagement and completion of that journey. With the collaboration and validation of others, moral satisfaction is obtainable—to a degree, but is limited by them in other respects. "Freedom" as an adjective which describes a "moral" state organizes around prescribed ranges of acceptable self-expression that communities of persons work on together, and in that respect, "freedom" is about becoming and choosing one's identity as part of a social commitment.

Such choosing appears to have continuity across cultures, though each culture seems to support and integrate its own version of it. There are defined ways, unique to a given social system, of becoming an "adult," of claiming rights of property, of creating circles of trust, and of generating respect. These are gained with difficulty, becoming part of one's identity, and one does not easily set them aside.

But something like this happens in the mind of one who makes the ethical transition from moral to political freedom—the topic under examination in the fourth essay. Political freedom is proof of sorts that the human mind is capable of moderating its response to challenges by diverse moral systems.

The capacity of human consciousness to organize governmental structures that accommodate and process the moral claims of diverse ethnic groups is no less of a modern marvel than scientific advances in the fields of particle physics and molecular biology. And a credible argument can be made that without such structures in place, there would be no place for science at all.

That structures of mind correspond in some way to institutional structures of government has long been part of the Platonic tradition in which a three tiered model of mind corresponds to a social and governmental aristocracy—a rational tier prevailing over physical cravings and social ambitions. Freudian and neo-Freudian models of the psyche also partition into a tripartite psyche, and place significant emphasis on the tendency of mind to cast its own structures into its various externalized creations. Thus we are tempted in the development of a theory of political freedom to contemplate the relationship between the structure of mind and human institutions.

I must admit that there are a number of things happening in this essay simultaneously: (1) a theory of mind which notes and explains agreement between institutional hierarchy and consciousness; (2) a theory of language designed to emphasize the interrelatedness between tiers of consciousness; (3) a theory of government which emphasizes the continuity and accountability of dialogue between sovereign and subject and; (4) a theory of conscience which empowers political discourse to be accommodating and receptive to diverse moral viewpoints. Together these come close to what most mean to say when they use the expression "political freedom"—government which is recognized by, interactive with, and protective of the morally autonomous communication of its citizens.

Political freedom might at times be pragmatically conceived in terms of the elimination of oppressions, but the avoidance of oppression is hardly a guiding principle from which one might devise technologies of free government. The attachment of "freedom" to politics suggests the affirmation of value, and it is probably fair to say that in the end I was able—as a consequence of an inquiry into the nature of language—to write more selectively about that value.

Some thoughtful readers have communicated to me that the introduction of the term "deity" to political discourse is unclear. I understand the concern. Part of the problem is that it is a supposition made from its effects, in much the same way that *reality* is a supposition derived from its effects. It is in that sense pre-cognitive, and whether it is conceived as a psycho-social influence, or something else is unknown. As a supposition, "deity" functions to balance a linguistic equation in a way that bestows political life on conversation. Like most assumptions, its value consists in the coherency it lends to a theory of language, and derivatively, a political theory.

I might have ended a treatise entitled On Freedom there, inasmuch as abstract assertions about the hierarchic structure of consciousness seem to lie close to the margins of acceptable theory, but then it would have been difficult to refer to this as a treatise on "freedom." It might then have better been called "on competence" in that scientific, intellectual, moral and political capabilities had merely been described in terms of their efficiency toward an objective—with no ultimate objective identified. Indeed, there is no appreciable difference between "requisite" and "competent" as both attach to the viability or efficiency of a given organizing consciousness.

A *requisite* discourse engages a world of ethical principles applicable to political systems—systems of interactive influence. Descriptors such as "trustful," "peaceful" and "rational" tend to predominate in ethical systems designed to govern political systems. And such a world has difficulty with descriptors such as "joy," "inspiration," "excitement" and "risk."

Escape from the tranquil ethical pose of being in control of matter is hard. We are born into a natural world of material substances and processes, and science informs us that our elevation from that world is not much more than the formation of progressively more complex and sophisticated sorting mechanisms. Does this change, however, if we admit to life after death (or before birth), the healing of sickness through prayer, and the termination of hatred toward deadly adversaries? Is there a "spiritual" reality, and if so, what relation does it bear to matter?

Theology—particularly abstract theology—is a pastime that can lead in many directions. Even so, bad theology can do much harm, and there is much of it in public circulation. A coherent theology ought to have a theory of atonement, where God's goodness is not marred by the various predations brought on by material competition between living beings. However one explains it, the real question is not whether the explanation makes sense, as much as whether one can receive and express spiritual love, an experience that many equate with "goodness"—an ethical principle experienced pre-cognitively.

We are, as a species, moved by this experience particularly when it appears in the least of us. It reminds us of something better than the attainment of dominance over others, and the perpetuation of our bloodlines—something spiritual which overwhelms the importance of material goals. That is essentially what freedom is in all contexts, an affirmation which dissolves boundary. A spiritual affirmation of value as "freedom" represents a point of orientation which a "requisite" platform has difficulty reaching. This orientation, it seems, affects everything having to do with human organization.

Take intellectual freedom, for example, the subject examined in my second essay. A significant intellectual paradox which afflicts the logic taught in college classrooms throughout the world is solved through the recognition that essential to the formation of objects is the intervention of value—of consciousness (spirit)—and that objects acquire form through that intervention. The recognition that no object can confer value on another object essentially solves what is known as Russell's paradox.

Freedom prompts us to wonder about what is valuable in the appropriation of a moral code. How does it reflect goodness? In what sense is existence affirmed by it, and how, by knowing its limits might we transcend them? Without an interest in freedom, such questions are not asked, and therefore not answered.

But asking such questions is part of the intellectual foundation for the development of a theory of political freedom—which examines systematically the importance and the limiting characteristics of tribal normative systems. We write and sanctify constitutions so that those systems will not sabotage the formation of nations. It is in fact spiritual receptivity to an ethical universe that allows us to be religiously tolerant, and appreciate the value of religious diversification.

"Spirituality" is not consciously present in all decisions we make about organization, but is, perhaps, in our most important decisions. The world is presently afflicted by considerable anxiety and suffering related to the acceleration of social, political and economic change, having religious implications for billions of individuals. Over the next fifty to one hundred years this suffering is apt to become much worse as competition over resources in limited supply—most noticeably water and energy—reach crisis.

There are two major courses this transition might follow: (1) a political course through which nation states compete violently for control of these resources, or (2) a cooperative interface between modern and pre-modern societies which effectively accelerates the process of political, economic and technological modernization. If we feel as strongly about freedom as we say we do, there is a fairly good chance that we will choose the second option.

Viewing this work now retrospectively, it seems that if we as a species feel a sense of triumph over the implementation of free academic, social and political systems, we are likely to be more successful in preserving the gains acquired in those systems—and in sharing them with others—if we understand more about what "freedom" means. All told, it is tougher to get one's arms around it than it may appear at first glance.

I should say a few words about the organization of this book. It follows what can be described as a narrative, in that it moves in stages, each of which combines more than one objective. This is evident, for example, in the first chapter on Scientific Freedom, which endeavors to situate "organization" within a philosophical tradition, the introduction of concepts that have moved organization to science, and the struggle of this science with ethical neutrality—all of which are more confusing to address when treated as separate topics. In an earlier attempt to break this manuscript into topic headings, my impression was that it subtracted from more than added to its clarity. If the narrative standing alone could not speak for itself, it probably needed to be reworked. Titles vary with the interest of the intended audience, and I concluded that this book might be of interest to more than one audience—specifically to one interested philosophically in freedom and another interested in a science of organization.

On Freedom: Organizational Science Examined Philosophically

It is difficult to say what it means to examine something philosophically, except to draw upon a philosophical tradition and endeavor to evaluate assumptions behind a given perspective with depth and honesty. Philosophical labeling, at least in my view of things, tends to divert one's critical faculties from the task at hand. In this book, it is about assisting a scientific dialogue with the development of some measure of clarity about what is known and what is not. If science assumes that the nature of things reveals an ethical void, that assumption is not effectively examined scientifically, but philosophically. In that sense, philosophical assessment is the principal mechanism by which the term "freedom" can be considered meaningful.

On Freedom:

Organizational Science Examined Philosophically

1.

Scientific Freedom

"A human being experiences himself, his thoughts and feelings as something separated from the rest, a kind of optical delusion of his consciousness. This delusion is a prison, restricting us to our personal desires and to affection for a few persons nearest to us. Our task must be to free ourselves by widening our circle of compassion to embrace all living creatures and the whole of nature in its beauty."

Einstein

Peter Gibson Friesen

1. Scientists may contend that human values have evolved through competitions arising in the wild, where the love extended to one's intimates promoted the survival of the social organisms in which we live. By this logic, one may argue that a love that extends to one's enemies is wasted. Yet this wasted love could be the love from which science and scientists are made.

Our ascendancy as a species is often attributed to an intellect that allows us to dominate our environment. Yet within the last few centuries this intellect has changed the world into something which is barely manageable. Thus, from an enlarged perspective humanity might appear to be nothing more than an aggressive strain of bacteria, which in the course of a few hundred generations, devours its environment and dies.

This image of the human condition dissolves upon contemplation. The simple fact that one may, within certain environments, be clever and skillful in addressing biological needs, is not a real point of distinction. Despite many blind tendencies, we care about who we are, and perhaps more importantly, what we are. This caring has united many in a dialogue on the human situation, and much dialogue on the power of science to reconcile humaniy to the world.

Science emerged in a European civilization heavily influenced by a religious movement which emphasized the intimacy and kindness of a relationship between a mind that created the universe and a human being. That movement was corrupted by political interests, and made a number of aggressive and erroneous claims about human nature and the Cosmos. At one point a consensus emerged that such assertions were untrustworthy.

Whether scientific freedom was somehow included in the spiritual orientation of the European mind depends largely on how one defines "science" and "freedom," among other things. It seems, however, that science and freedom did at some point become dominant themes in European civilization, and its colonial extensions. They reflected a significant change of perspective—an epiphany which continues to gather momentum throughout the world.

While it is good to remove political interference from scientific investigation, it is naïve to suppose that "freedom" consists in the removal of such restriction. The human quest for freedom is marked by the irony that many of those who claim to have been liberated from a given oppressor recreate the same oppression elsewhere. It is not clear that the removal of political limitations and boundaries precede the establishment of freedom, but to the contrary, when freedom is achieved, those boundaries begin to yield.

The encouragement of freedom by government is good for those who would otherwise suffer oppression, but it confers other benefits which are less visible. As individuals progress toward freedom, they become better friends. They become more trustworthy and honorable, and look for the same in others. They are kind and joyful and affect others contagiously. They will gather with others to claim ownership over a government which tolerates collectives and encourages personal initiative. They engage in science.

Freedom is thus a cause rather than consequence of science, allowing science to be what it was meant to be. But what is freedom? Is it something that exists within the mind, or is it more correct to say that freedom is the soul from which mind extends?

Scientific freedom is manifest in the periodic reorientation of thought toward the world. A "thought" identifies a place within a universe. A thought can be simple or complex, or both simple and complex. Scientific freedom does not address, as some propose, the way in which thoughts are formed, but addresses a value which precedes the formation of thoughts. It is a state of confidence joined with caution which supports investigation and theoretical revision. Science is the result of this value.

This simple idea is incompatible with the notion that transformative events in science are the product of observation alone, or that the advancement of science is a gradual and natural consequence of the interaction between an intellect and the environment. While human rationality may have always existed, science did not. Science as it is known in "modern" civilizations occurred because of a shift of human self-awareness. This change continues in the form of more unified theories of nature and of human behavior.

As a natural universe appears to be yielding to the scientific method, a moral and political universe resists it. The movement within the behavioral sciences to reduce human consciousness to natural imperatives, a collection of motives oriented to the genetic survival and propagation of a human species, arouses discomfort in those who insist that the human mind is in some sense elevated above animals. Can one maintain a scientific fascination with nature while affirming its freedom? Does an expanding natural science diminish its importance or simply alter it?

According to the Oxford dictionary, the root of the word "free" is the Sanskrit "pri" translating as "joy." This later became "freon" in Old English meaning simply "love." It has the same root as the word "friend." Freedom of this sort is quite different than the absence of restriction. It has positive content, involving the presence of value and cannot be defined by the absence of restriction alone.

Instead of the word "freedom" it may be less confusing to use the words "love" or "soul," which seem to more directly denote the feature of consciousness that produces a "free" individual. However, that approach may do more to obfuscate than clarify the nature of freedom, in that freedom is not a consequence of love, but the affirmation of it. Freedom is not merely descriptive of the escape from conceptual, emotional or social confinements, but a living imperative that these yield in the presence of value.

Yet at this stage of the discourse, such assertion regarding the union of freedom and love is hyperbole, and it is not clear what, if anything, love has to do with science except, perhaps, that it surrounds and influences mentally imposed boundaries from which our sense of reality emerges.

2. Science places emphasis on rules through which we attribute reality to objects, and the order through which they appear to us. Knowing that the mass of one object will affect the motion of another, and that it will do so consistently as one places two or more objects together in a given space, amounts to a scientific accomplishment.

Something more happens, however. As one conceives of a rule or pattern of interaction between objects, he or she supposes that there is a "force" or "power" governing their relationship. Therefore, to suppose that mass is associated with something called "gravity" or with another thing called "energy" is a way of ascribing content to objects inferred from but not sensed in their appearances.

This is part of what it is to say that something is "real," i.e., that it has a power to affect what one perceives independently from what one may otherwise desire or anticipate. That certain objects can satisfy hunger, conceal other objects, or serve to reflect light, all suggest that there is reality beneath appearance. If we were to sit at a feast and eat without ever feeling nourished, we may suppose that the food was not real as it lacked the power to affect our hunger. The experience of a "reality" existing apart but only inferred from appearances is what is properly labeled "scientific freedom." It is a way of seeing beyond appearances.

While there is much academic discourse on what is properly included in a scientific "method," the structure and validation of this method belongs to a theory of knowledge or "logic." One who violates rules associated with this structure is apt to encounter the criticism: "that is not science." There is a difference, however, between the rules constituting a scientific method and the mental orientation that places value on such rules. Accordingly, science begins with confidence or faith in the proposition that there is a reality "out there" and that this reality is knowable. It is not uncommon for the mental void of an exterior reality to be filled with sanctified notions organized by religious or semi-religious systems. The will to understand the world scientifically must be great indeed to countervail against such religiously sanctioned constructs.

Where is it written that reality must make sense to the human mind? Can a religiously cherished explanation of reality be overthrown by anything less than an inspired commitment to the knowability of reality? Why do we suppose that we can "know" reality, except by supposing that we share something with it?

Consider a hypothetical divide between the human mind and a reality independent of that mind.

$$(\text{Internal}) \; \text{External}$$

Regardless of the persuasiveness of evidence for a given theory, there is risk associated with one's attachment to the theory. Every question about reality involves assumptions about its knowability, and as a consequence mind itself often becomes the subject of investigation. Thus an altered diagram: [1]

$$(\text{Internal}) \; \text{External} < \begin{matrix} \text{mind} \\ \text{matter} \end{matrix}$$

One may suppose that a science of matter represents a different kind of science than a science of mind in that one has direct experience with mind, but not with matter. However a distinction such as this misses the real point, which is that the entirety of the diagram occurs within mind. A distinction between internal and external, and a reality in which a distinction between matter and mind appear, are both mental inventions. The distinctions are a useful way of positioning mind so that it may respond to influences over which it has no control.

Those influences may be quite different than anyone imagines them to be, and the development of an understanding of them is the task of science. The application of the term "freedom" to science is a way of emphasizing a state of self-awareness that is both confident in the existence of a cohesive reality and willing to adjust to error.

As science progresses, "matter" and "mind" are both subject to redefinition. Such redefinition becomes more important as one encounters situations where value affects not only the appearance, but the content of reality. That science in its present state emphasizes the dependency of mind on matter tends to beg a number of questions, not the least of which is whether complete acquiescence to the confines of matter is a good thing, and whether science involves a demonstration of freedom from such confinement.

How does one comprehend freedom? It is difficult to say what it is that comprehends it. To say it is a "self" is confusing. The status of *self* changes while freedom is more like a constant to which one evolves. The *self* emerges into the awareness that freedom is what it is, and not an attribute which it obtains. It is a state of realization, and not of acquisition. Freedom uses the user as he or she contemplates it, and requires that they become what it is they seek to understand. That is to say, if the supposition of the inquiry is true, the inquiry is biased.

In that sense, there is no separating a regard for the ethical substance of mind from a celebration of the cohesiveness of nature. Belief in a comprehensible universe supposes that the human mind is in some sense fulfilled or satisfied through the investigation of that universe, as a participant in the "mind" of that universe. Even as one devises theories that diminish the importance of the human organism, the fact that one is engaged in such a task reflects entitlement, which treats ignorance as an ethical void that must be filled.

Altered views of nature are thus apt to set off tremors in a given ethical landscape. This was the case as scientists came to view genetic survival as being as important to animal behavior as gravity is to celestial movement. An analysis of mind then became focused on cognitive events because cognition stood out as our principal adaptive faculty. Ethical and metaphysical events, being more removed from the survival of the human organism, are regarded sterile, the frolic of speculative philosophy.

Despite our best efforts to secure ethical neutrality in the evaluation of an inanimate reality, these efforts falter even more as our gaze shifts to the sentient reality of human behavior—the reality in which the organization of mind and social relations occurs. The attempts to restore ethical neutrality to the study of mind have in recent years focused on the neural mechanisms of the brain, leaving very little room at the table for ethical philosophers. Part of the problem with this approach is that science is not, and never has been, ethically neutral.

3. Near the beginning, there was no line of distinction between science and ethical philosophy. Ancient thinkers took a more balanced interest in the nature of a physical and a mental universe, and devised theories emphasizing equivalence between our ethical nature and the laws which governed our material surroundings. By modern accounts, the human has existed in its present state of biological evolution for at least one hundred thousand years. Given the superstitious nature of human reckoning over these millennia, the emergence of philosophy, and what later came to be known as "science," was a fairly sudden event.

Maybe it was inevitable, though there are circumstances in which the occurrence of science is less likely. In societies where social organization is not complex there may be insufficient debate and discourse for the emergence of conceptual systems suitably labeled as "scientific." This is not to say that these societies are unsophisticated, but that there is too little interest in a socially validated dialogue apt to evolve toward a science, or even a philosophy.

Assuming that a complex and hierarchically organized society exists to support the evolution of sophisticated languages, a society could fail to develop science for other reasons, such as a protective and fearful government intolerant of concepts that tend to undermine it. It is difficult to imagine the development of credible and comprehensive theories without a long process of challenge and reformulation—unwelcome in societies under totalitarian government.

Where philosophical thinking is most likely to occur is in societies whose political relations are intricate enough to favor and encourage abstract deliberation over human nature, and in societies whose government is secure enough not to be threatened by it. Particularly where disputes are resolved through the process of debate and persuasion, and social value is placed upon that process, many will take the trouble to contemplate what a human is, and express their views publicly.

That human nature may be something different than what socially reinforced institutions say it is may, at points in history, attract debate over what that nature is. Debate is apt to focus on what, if anything, distinguishes human from animal. Where thought touches down in such debate depends significantly upon one's ethical orientation to the world, which includes how one manages one's self and their relationship with others around them. Such orientation is apt to affect the kind of government they choose, and those who they would have serve as the representatives of that government.

The interplay between political government and self-government was a matter examined long ago in ancient Greece in the work of Plato. In a work he called "The Republic," Plato began by considering the proposition that justice is the interest of the stronger. Socrates refuted this contention offered by an irritating student by pointing out that justice and its opposite, injustice, were under that proposition, the same thing, and that the proposition was the equivalent of saying there was no such thing as justice.

Although Socrates had subdued this challenger, another of his students pointed out that he had dodged the more important concern of whether justice existed in any meaningful way—such that one might recommend it to another. The student posed this question. What if there was a ring which a person could wear which would make that person invisible, such that he could take whatever he wanted without getting caught and punished? Why would it be in the best interest of that person to behave justly?

The question could not be answered without a discourse on the nature of human consciousness. Socrates argued that consciousness is multi-faceted, and can be governed by any one of three elements in its nature. He recognized that it possesses animal appetites, and that an "appetitive" element drives it to satisfy its passions and desires. He also recognized that human consciousness cares about its relationship with others, and thus about its status and stature. This he referred to as the "spirited" element. He further recognized that consciousness thinks, and that this propensity has a will of its own, and thus exists as a "rational" element.

According to Plato, human happiness is dependent upon whether or not it is ruled by the third element, a rational element, and that only through this rule can harmony between the competing interests of appetites and social vanities be established. In a sense, this analysis begs the ultimate question, because it would seem that harmonious rule over one person's desires may yet pose conflict with the same kind of rule in another person, and that one person's justice may result in injustice to another.

But this is exactly where Plato reached a point of inspiration. He met this objection by arguing that this rational element seeks harmony universally, and not just a harmony of desires within a given person. An individual ruled by his or her rational element would choose to extend beyond themselves, and to structure society and its government in such a way that a harmony of interests prevailed over disharmony. Plato then offered his "republic" as an utopist vision of what this rational element would choose.

That was his answer to the question—i.e., doing evil violates human nature and would leave it unfulfilled. There is so much elegance in this response to the question that it might be better to leave it alone without criticism. However there is significant uncertainty over whether this so called "rational element," if indeed it does make a demand of its own, would make a demand of sufficient strength to compel one to act justly in the face of violent resistance.

Plato's theory recognizes a workable symmetry between an organizational hierarchy within the mind and one that is the result of collective action—socially formed hierarchical organization. Not surprisingly, the role held by the "rational" element sits on top of this hierarchy and manages subordinate roles into orderliness through mental skills focused on the anticipation of the behavior of others. In this sense, the study of consciousness and the study of organization are one and the same. Disorder emerges from time to time, but is identified with a departure from reality. This was among Greek philosophers proof, in a way, that order and "reality" are equivalents—at least in the examination of social constructs.

The elevation of human intellect to a position of dominance may be a necessary condition of scientific freedom, but is it a sufficient condition? The Platonic suggestion that there is a fit between human nature and that of the world it inhabits, and that happiness can be advanced with the rational discernment of this fit, thus authorizes and empowers one to search out the nature of all things. Such confidence might have brought the ancient world to the threshold of scientific freedom, but it took some time for humanity to fully appreciate that the celebration of intellect has a number of undesirable consequences.

The hubris of a dominating intellect is apt to be attached to its creations—sometimes referred to as theories. The intellect which forms these theories often functions from a dull awareness of what motivates it, and may thus be driven by truly craven motives. The sense of entitlement of a dominating intellect committed to a primitive view of man may compel savage deeds. Ingenius behavior is apt to be ingeniously savage. And science would—and did—become stymied by the political need to control the lives of others.

This encounter may have prompted religious consciousness to alter the moral question of Plato's Republic. Instead of asking why be good when one can get away with evil, the question became, why act justly when the perceived cost of doing so is grave, as in loving one's enemies? Some believe that in answering this question, a way to heaven opens that emphasizes the value of existence, a value which frees the mind, and along with that, produces a spiritual willingness to endure persecution. Partnered to this belief is a fascination with the narrative of the life of religious figures and the establishment of religions that idealized those lives.

4. From another vantage point one might argue that regardless of what religions did with the spiritual messages emerging from the axial age, the die was cast irrevocably toward science. Believing that the universe was and is created and activated by a singular, complete and omnipotent consciousness—having an intimate and benevolent relationship with humanity—includes a belief that the universe is knowable. One is less inclined to labor toward the resolution of perceived conflicts between theory and evidence in a universe or pantheon of competing deities, or in a world of one or more predatory devils. Why waste time trying to make sense of a world which is not supposed to make sense?

Within a culture that embraces "love" as the activating principle of all that really is—as "God"—one is apt to find individuals with unusual gifts of imagination seeking a unifying theory of existence. Political rulers have made whatever use they can of such theories to control the behavior of their subjects, more often than not distorting the meaning of them. Such was the case in medieval Europe. Accordingly, about five hundred years ago the spiritual supposition that we are kindred to God emboldened resistance to the venality of religions created for secular leaders. As a consequence, thought strayed from the sanctification of religious institutions, while holding to the sanctity of human consciousness. Religions reformed, and science emerged as part of an ethical affirmation of the value of humanity.

There are many who view science as a complete separation from the religious inspiration which prompted the formation of religious institutions, where skepticism replaced faith as the orientational principle by which one makes sense of the cosmos. This view is simplistic and tends to ignore the words of scientists who speak of discovering the mind of God, even as they admit they cannot define what God is. Of course, the reference to God may be no more than a metaphor, but the point of the metaphor is to emphasize the fervor with which science is practiced.

If the human organism has existed in its present state of biological evolution for more than one hundred thousand years, why has science emerged only within the last five hundred years? One could say that we became smarter than we were before, but that is unlikely. The hunters of primitive societies had to be very smart to survive, perhaps even smarter than we are. It likely has nothing to do with intelligence. Missing from the primitive mind was an ethical commitment to "order" as a knowable feature of a coherent universe—the value from which science emerged.

Plato's vision of a world functioning under the rational imperative of harmony reflected his vision of a human consciousness subject to perfect organization by its rational faculty. The human expectation of a natural order within the human mind was a way of participating in a perfect order manifest in the universe. Efforts toward its discernment would be rewarded, and humanity might emerge from a cave of illusions, to the direct apprehension of the principles which governed all. In this way, he promised scientific freedom along with happiness as a benefit of obedience to one's rational faculty. Not surprisingly, the Platonic expectancy was never limited by an interest in social organization, and extended—rightly or wrongly—into the cosmos.

Yet Plato, along with a number of other ancient philosophers, was overconfident in both the power of reasoning and the availability of nature. They did not achieve science or develop what is commonly known as the scientific method. At some point later, much of humanity placed its mind in a proximate relationship with a mind in which and through which existence obtains value—imagined by many as a God in a state of "love" with humanity. It was the kind of thing that might encourage a scientist to toil over the nature of things just for the sake of knowing, so much so that one might attribute to it a point of origination for the science we are familiar with in the modern world.

However this attitude—wrongly regarded by some as uniquely Christian—tends to attract a scientific argument against the existence of God. A scientist examining a very small human, compared to a very large God might choose a simple explanation for the commitment of men and women to God's existence, one that discards God's actual existence as a necessary aspect of the human consciousness. Human vanity is the real authority, so expressed through human intellect.

Pursuing this hypothesis, one may then argue that a divinity representing an ideal being is an extrapolation of human consciousness. A deity supposed to exist independently of the subjective experience of a given organism is a fictionally projected self-image representing an idealized compilation of human virtues. God is thus a way of personifying the attributes of human consciousness we find useful in our practical affairs—and thereby introducing an ethical dimension to a world which is, in reality, ethically neutral.

Why wouldn't they reach this conclusion? The expectancy that a transcendent mind extends in some meaningful way into nature comes with the disappointing realization that nature seems to respond with indifference. More than that, survival seems to have much to do with one's ability to defend against nature. And so, this "mind" we think of as creator of an ethical universe is a fabrication, a survival strategy oriented to the formation of cohesive social systems. There is but one ethical principle: survival and procreation. If one asks by what authority this conclusion is made, the answer is philosophically uninformative. "Reason" is the authority.

Those impressed with the functional utility of human intellect might imagine a perfectly functional mind existing independently of any given intellect, and call it the "reason" of God—the "reason" which makes science possible. Within an analysis of this sort, even the reason embraced by scientists can be viewed as an unnecessary abstraction applied to our organization and use of information. Why create a repository for reason, in other words, and call it "God" or why would cognition, for that matter, repose in an abstraction called "reason"?

The error of this argument does not consist in the point it makes, but in its claim of final resolution. The projection of human limitations into an object deity does not prove the nonexistence of God, but only that the source of the projected limitation is itself constrained by limitation. An argument for the existence of God is another matter entirely, having much more to do with the fact we are often impressed with how stupid we can be, and that we are often drawn by surprise into elegant and moving visions of reality. That order is, in a strong sense, discovered rather than manufactured is the basis for what has historically been referred to as the "ontological" proof of the existence of God—something we shall address later in the book. Within this tradition, God and reason are more like enigmatic constants which we comprehend in very small increments.

The historical defects of science are not the result of a belief in God—whatever that is, but in the endeavor to make God into an extension of man. It is more likely that human thought, tasked to understand reality (the mind of God) instead of making it into something easy to know will address its own theories skeptically. We may gaze upon the universe with the expectancy that we will come to know it because we share something with a principle of order that accounts for it. Skepticism does not compel us to truth as much as it prevents our sense of truth from being congested with superficial theories.

The risk of congestion is larger in the behavioral sciences, where the universe under examination is made up of interests coded in myth and metaphor. Here it is even more important to quiet the noise—often dressed in the costume of "philosophy"—in order to enable the occurrence of what we think of as "science." How does one do this given that human behavior makes sense only after allowing for the existence of very moving human affirmations?

5. Even the worst problems of science—which are correctable—hardly detract from the historically proven effectiveness of it. As the power of science came to prove itself as a mentally superior attitude or orientation toward the world, governments that had once sought to prevent it from happening realized that they needed it to provide for their subjects, and to protect themselves against rivals.

Even with the liberation of science from religious institutions there remained the unsettled question of whether a scientific analysis of the human tended to elevate or debase our self-image. The so called "scientific revolution" thus marked the beginning of a debate. At one pole of this debate are those who emphasize the effectiveness of human intellect as an adaptive faculty, allowing one to handle material challenges successfully. At the other pole of this debate are those who are impressed with the capacity of reason to disengage from matter and devise elegant theories to comprehend it.

The first of these is frequently referred to as "empiricism" and tends to lend priority to the natural world. Empiricists propose that reason is the imprint of that world upon consciousness. Since the world is governed by laws of causation and substance, reason is a formatted adaptive mechanism that is sensitive to those laws.

This way of thinking about reason appeals to one of the principal rules of science often referred to as "Occam's razor," which was offered by William of Occam, a 13th century philosopher. This rule requires acceptance of the simplest and most ordinary explanation of events when two or more explanations are offered for a given set of facts. To many scientists this razor favors the elimination of a distinction between man and animal, and this tends to rule out a special relationship between the human organism and a "mind" which made the universe.

The second of these, which is often referred to as "idealism," emphasizes that all events associated with what one calls "the world" occur in consciousness as ideas. That world would be a void, except that it presents many surprises, and resists us. Yet despite this, we have as a species experienced some impressive breakthroughs in our comprehension of this world, and as we reflect upon our emergence from more primitive states, we sometimes hope that we exist in a special relationship with a "maker" of this world—as a mind contained within and nourished by a surrounding mind.

This, it seems, has been and continues to be a paradox of science. Science is motivated by an ethically vibrant imperative to understand, while deploying within its discipline assumptions and attitudes which demoralize this imperative. This paradox did, indeed, account for the philosophical schism in a period of history referred to as "The Enlightenment" until a point in the 18th Century when a German philosopher, Immanuel Kant, attempted to resolve the conflict.

Kant recognized that the world as we construe it is reactive to both interior and exterior influences—one which impresses our physical senses and another which resolves them into unified wholes or objects. The latter of these two influences he called "reason" the properties of which could be educed through what he referred to as "a priori" reasoning. This kind of reasoning examines what is necessary to the formation of objects in the world with which we interact. One educes "space" and "time," or "causation" and "substance" as properties of reason "a priori" if one cannot imagine a world without these properties.

According to Kant, reason includes an imperative of unity, i.e., that the world, though presenting what appear to be contradictions and inconsistencies, is in fact consistent when reason understands it. He called this the "transcendental unity of apperception," which alludes to the transcendent property of reason to resolve the world into unified theories. This imperative is thereby embedded in all that exists.

This view is different from what Plato referred to as the "Rational Element" in human nature, which is a human faculty of organization.

For Kant, reason does more than organize objects presented by the world. The unity contained within every object or thing is the product of reason, and governed by an imperative that is active with or without human desire, and is in that sense "transcendent."

While there is some risk of oversimplification here, it seems that part of the ancient model of consciousness is a rationality designed to comprehend a formally organized universe.

For Kant, reason's transcendence consists in its persistent role in the formation of organized constructs (objects). This presents a more complex picture in which organization itself is the product of the interaction between rationality and an exterior reality, and not simply an ordered domain that one's rational faculty discerns. Organization is more than what we do with objects, but is about how objects come into existence.

Under this model, the properties of reason are educed "a priori" by determining what features are necessary to the formation of coherent mental constructs. They resemble the structure of mind because mind structured them. What we think of as an "objective" or "object" experience is descriptive of a synthesis between an internal influence, mind, and an external influence. The scientific mind is one that seeks agreement between these influences, and thus engages in a process of self-examination and adjustment, without which science tends to be imprisoned within dogmas. An unreflective and overconfident mind tends to impose rather than discover order.

One can hardly deny that Plato—and other Greek philosophers—were hinting at something like this all along. In both argument and allegory, they promoted the notion that the universe is bound together by a principle of order ("form") that can be discerned through contemplation. What changed was where "form" was located—from somewhere "out there" standing by to be noticed to somewhere "in here" as an activating source of order, a set of organizing propensities. For Kant, "form" is a set of constructive principles, accompanied by an imperative seeking unity and consistency. These influences precede cognitive output, so much so that all that we can imagine as "existence" depends on them.

Accordingly, science is a creative process of adaptation between theoretical assertion and factually based observation. Science is a way of treating "order" as a supposition bound to observation that cannot freely progress without valuing the relationship between supposition and observation. Rather than being disturbed or disrupted by disagreement between theory and fact, science is liberated by it.

So, at last philosophical thinking had matured to a point where one could imagine that the existence of the world might depend on a presence—call it the god of "reason"—which emphasized the direct participation of the human mind in the discernment of order. It was, however, a participatory process governed by standards allowing us to distinguish better from worse in a community of scientists—an ethical process.

Would the god on whom we base existence be a god for whom one might endure persecution? It seems that Kant implied as much in his moral philosophy. For Kant, reason is no less important to the human in making sense of itself than it is in making sense of sensory data from a presumably material reality. Reason's influence on the formation and validation of human interests constitutes a moral imperative.

Thus Kant addressed the question posed by Socrates two thousand years before, i.e., why be just? For Kant it is not the fact that we possess a rational desire which seeks satisfaction—which was Plato's solution to the problem. In fact, Kant renounces human affection of any kind as the basis of a moral philosophy, and instead views reason as the condition precedent for meaningful action. It isn't a happier existence that he advocates, but existence itself, the basis of which consists in reason.

Kant's moral rule requires that action be such that if a maxim or rule of conduct were made of that action, it can be made universal law. He insists that this imperative, when ignored or disregarded, results in the dissolution of the very ideas that define one as a being. For example, if one engages in theft, the idea of "property" dissolves in mind, only to be replaced by power over objects. If one engages in adultery, the idea of "marriage" loses existence, and is replaced by power over a relationship. If one engages in murder, the idea of "life" is lost, and becomes power over survival. If one lies, the idea of "truth" is disestablished in thought, and becomes a matter of rhetoric. If one covets, the idea of "self" weakens, and yields to a more vacuous sense of social comparisons.

What the world retains from his moral philosophy is the provocative idea that moral judgments arise from something acting independently of human desire. The same feature of consciousness which makes one a scientific being makes them a moral being. People refrain from theft for the same reason they would resist the aggressive assertion by a social constituency that two plus two equals five.

Part of the difficulty with this idea is the simple fact that reason as it is popularly conceived fails to evoke feelings of moral inspiration. There are no tears of commitment—or heroism and self-sacrifice—in reason, nor does the rational deduction of a moral principle resonate with one's experience of the "goodness" of a moral act. Even science, which prides itself on the evaluation of reality dispassionately, is more inspirationally rewarded in its insights than one would be under the Kantian model of the moral human.

As with all great philosophers, if one is patient and charitable in their reading of Kant, it is possible to extract an intuitive vision of moral and natural universes which are in some respects interchangeable. The cohesiveness of reality seems to consist in the rightness of it, while the rightness of moral behavior consists in the cohesiveness of reality. Are we saying, then, that nature has an ethical pulse to it, or the other way around, that ethics replicates nature?

In a roundabout way, the philosopher's world had come upon the same impediment found in ancient Greece—a unified theory of ethics and epistemology which leaves consciousness groping for meaning. Here there seemed to be a similar—though less obvious—overconfidence in the power of intellect which afflicted the ancients. And, not unlike the Romans, European nations began to claim rights of expansion due in significant part to intellectual hubris and the strength of the resultant technologies of science.

6. As much as a philosophical mind is comforted by clear distinctions between epistemic and ethical concerns, there is no true comfort in such a distinction. The Kantian notion of science is prescriptive, and not neutral. Science is inspired by a "truth" which stands opposed to error, a truth that eventually leads to introspection and to the organization of a coherent moral system. But the use of a crabbed notion of "reason" as a neutral instrumentality is awkward.

Accordingly, philosophy following Kant has proceeded in two distinct though related directions. One of these is to leave "value" out of the subject matter of thought, and to focus on the identification and definition of the characteristics of reason, whose principal function is to formulate true judgments. Reason also combines propositions into compound propositions which can be evaluated for truth. Since these compound propositions can be rather complex and sophisticated there is much for logicians and mathematicians to contemplate.

Yet as much as logic progressed in clarifying rules of order in a "science" of reason, something was missing. Alfred North Whitehead, a collaborator with Bertrand Russel in the logical treatise entitled Principia Mathematica (regarded the most important work on Logic of the twentieth century) challenged a propensity among scientists to trivialize mind by reducing it to material structures. Ludwig Wittgenstein—Russell's most famous student—likewise emphasized that language and the necessity we attach to assertions within language emerge from a socially active environment. Mind, along with its creations, has its own life and rules, and takes care of itself.

These philosophical objections were certainly around before Whitehead and Wittgenstein commented on the science of mind. In some beautiful philosophical works, notable German philosophers such as G.W.F. Hegel and Karl Marx have focused on the formation of human identity. Moral and political choices are inseparably connected to the social and political consciousness within which a person abides. Others, such as Friedrich Nietzsche and Soren Kierkegaard, viewed the fact of the dependency of the human "identity" on socially formed mental constructs as paradoxical given a human imperative and need to distinguish one's self from the herd. The "God" that had for millenia been regarded as the ground of human meaning was being marginalized by science and contemporary political systems organized around technological advances.

Accordingly, by the turn of the twentieth century, science had already begun to assert itself into the vortex of differences occasioned by the rapid advance of industrial civilization. This occurred in the work of Sigmund Freud. Since motives explain behavior, a behavioral science was for Freud a science of motivation. He was probably the first to formally acknowledge a distinct feature of consciousness which influences motives, but which is hidden from conscious recall. His "science" embraced investigative techniques by which such things that are hidden may be discovered. Then in a template of motivation which has survived in modified form to the present, he divided human consciousness into parts.

Underlying this template is conscious recognition of the moral paradox which arose in the work of Plato, i.e., the conflict between personal appetitive gratification and social expectation. These present themselves in consciousness in what Freudian psychologists later referred to as "id" and "superego." The libido contains, among other things, a self-assertive sex drive, while the superego collects moralistic limits suggested by one's parents. There is little emphasis on a supposed natural harmony between these sources of influence. Such harmony is asserted through the active intervention of a third feature of consciousness, the "ego," which mediates between the "id' and the "superego."

While it is easy to imagine that controversy attaches to the genesis and composition of these strata of mind, one is apt to be impressed with the fact that there are three, and that each is dominated by its own motive. This is, of course, similar to the model of consciousness adopted by Plato more than 2,000 years earlier.

An "appetitive element" would be the Freudian "id." That the "superego" replaced Plato's "spirited element" may not seem quite as apparent, except for the fact that Freud viewed moral rules as social constraints upon individual consciousness, and that a person's social status, which motivates Plato's "spirited element," is a matter of conformity to and fulfillment of socially imposed values. "Ego" resembles Plato's rational element, except that where Plato viewed rationality as a sort of project manager over the chaos of lesser impulses, personal and social, Freud viewed rationality" as a mediator in a primordial conflict between personal impulses and social restrictions. Human rationality plays a major role in that mediation.

In effect, Freud adopted Plato's tripartite model of consciousness, but replaced Plato's hierarchical concept with that of a dialectic in which irreconcilable opposites were synthesized by "ego" into a complex of compromise positions, often referred to collectively as "self." In this way, science added self-existence as a creative effect of reason—a "mind" in which we live and move and have our being. Accordingly, despite having modified Freud's sexually based theories of motivation, contemporary psychology is highly focused on the cognitive processes out of which healthy egos organize within a vortex of competing personal and social mandates.

One can look at the similarity of these models and conclude that science has made little progress in the study of human behavior over the philosophy of the ancients. Or one may, as a scientist, ascribe importance to this persistent intuitive recognition of a stratified mind organized within prescriptive boundaries.

As Freud acknowledged, the navigation of the psyche is largely a matter of constructing viable metaphors. His way of thinking, however, marked a point of departure from grandiose attempts in the past to establish (as other great philosophers) metaphysical continuity between mind and nature. Freud's mind was a struggle against the world, carving its identity out of a natural and social surround. Human ethical selections are naught but adaptive fabrications.

One might argue that the truth of a theory consists in its predictive value. Though plainly stated, it is not a satisfying argument. As it concerns the human organism, one may, by predicting, influence it. Thus depending on how strong such an influence is, one will seek to compare people acting within one context with others acting within another. At some point what is "true" will be attached to a type to which one attaches success. Behavioral science will always differ from physical science in this respect, i.e., one seeks to define and achieve an optimum, while the other is satisfied with accurate predictions.

Yet Freud's theories were regarded as a scientific advance. On the one hand, he identified places within the domain of human consciousness which, though not thoroughly understood, responded to stimuli presented by a therapist which suggested, nonetheless, they were there—a "reality" of forces existing beneath or beyond appearances. Such was the import of Freud's "subconscious," the place of repressed memories and experiences.

On the other hand, this new science was laden with normative judgments. A new behavioral science referred to as "psychology" aimed at the diagnosis and treatment of unhealthy psyches, emerged from the formal acknowledgment that factors or "facts" not yet considered could explain behavior. In fact, it was the existence of bizarre, "irrational" and severely maladaptive behavior which attracted thought to the possibility that the mind was like a physical organism, i.e., consisting of functional components which work together to support its viability.

The science of living organisms thereby sets itself apart from the physical sciences in its emphasis on the viability of such organisms—evident in the contrast of painful and maladaptive states of being to normal states, and their reform into more viable assemblies.

While there is a resemblance between modern psychology and the theory of mind advanced by the ancients, there is a major difference. In modern psychology, the phenomenon of maladaptation is descriptive of organizational failure—but this time not so much as the failure to seat rationality as an effective system manager. Rather, much of what we think of as maladaptive behavior is a consequence of failing to acknowledge and integrate the reality of disruptive life experiences, many of which are emotionally terrifying or heartbreaking. Rationality can oppress, suppress or distort reality.

The phenomenon of healing tends to refute the assertion that health is a relative value, and that "good" is only a socially sponsored norm. What attracted the attention of the scientific community to Freud's theories was the alterative effect of "psychoanalysis" on a variety of mental disorders. This process challenges a subject with an unhealthy ego to examine the assumptions about his or her life that contribute to the formation of the ego, particularly those assumptions formed through repressed memories. Frequently, when those memories are brought to conscious awareness, one learns that those assumptions are untrue, and the ego reforms to a state of health.

That health might follow after or accompany the exposure of falsity suggests that there is an ethical relationship between truth and health which may, in the behavioral sciences, be underappreciated. Does healing consist only of a sense of ease or liberation from ideas which had perplexed and confined the psyche within unmanageable contours, or does it consist in the affirmation of value of sorts which had been stifled or suppressed by error? Does freedom consist in the absence of restriction only, or does it include the presence of something else?

Effective psychoanalysts place importance on a phenomenon in the psychoanalytic process known as "transference." In this process the patient transfers the seat of self-judgment to the therapist, and allows the therapist to affirm the patient's value. Their role is to foster the patient's realization of "self-esteem." There is uncertainty among psychotherapists as to the real meaning of this term.

For this and other reasons therapy has shifted more and more toward a direct training of one's cognitive faculties, i.e., to examine the world with the objective of developing a clear and unimpeded perception of it, and to gradually increase the effectiveness of the management of environmental and social challenges. All of this comes with the assumption that "ego" is a formation of mind, and that "health" is to a significant extent learned. Yet there is great room for discussion as to the ultimate reward for the successful formation of ego. Is it a confidence induced relaxation of stress, or is there a value to existence released or liberated by a well formed ego? Is it the same sort of thing which religious thinkers describe when they use the terms "enlightenment" or "peace"?

This uncertainty comes with the territory, as there is an air of confusion and controversy prevailing over the entire field. This includes "self-esteem"—or rather the value of one's existence—and seems to represent conflicting orientations. One views human behavior as not more than the behavior of a complex animal struggling within the confines of the law of natural selection. The "science" of human behavior is essentially neurological—matter adapting to matter. Another views human behavior as something more, and yet stammers over the use of reason and "self-awareness" as the foundation for what that is.

In either case we examine something called a "mind" expressing preferences, and thus a universe coded with varieties of ethical selection— a mind which fills an ethical void with content and which manifests self-awareness in degrees. At some level this awareness reaches a point from which to ask "was I meant to know?" or "does knowing dignify or uplift me in some way?" In the behavioral sciences we seem to think that it does, even though we find resemblances between ourselves and a world of creatures that—unlike ourselves—worry very little about what they are. And this distinguishes us, in a way, i.e., that we fret over such things and they don't, and that our esteem is in some sense related to our understanding of ourselves and of the world around us.

7. No less important than a theory of "health" applicable to a human psyche is the shape and content of the human creations of that psyche. It was important to Plato—at least in his version of a "Republic"—that the elements of the human mind were replicated socially in personality types—persons driven by appetites, persons driven by social approval, and persons driven by their rationality. Indeed, the "Republic" sought to harness what he regarded as truths about a social organism which, in effect, replicated the structure of the human mind.

The significance of relationships of this nature was not lost on contemporary scientists of human behavior, particularly those trained in the psychoanalytic movement first advanced by Freud. In the analytically significant phenomena referred to as "projection," human internal structures are part of all external formations, and additionally, our sense of what motivates others "projects" from our own motives.

This, it seems, remains Kant's singularly great contribution to the development of philosophical thought, i.e., that the properties of mind impose an ineradicable signature on its creation. This influence is so intimate and pervasive that one can hardly refrain from calling it a deity, of sorts, as it lingers in and activates that creation. The scientist thus prepared to examine the creative effect of the psyche on the formation of objects is therefore likely to see constancy in the structures that emerge. There will be a noticeable similarity in the way simple and complex objects are formed. In fact, one can even study the psyche—in a manner of speaking—by studying its creations.

One therefore might consider whether there are "creations" that reveal something about a behavioral "reality" existing beneath appearances. Are there facts out there that can in some sense liberate us from the swamp of speculative philosophy and toward a science of mind? The clarification of the universe in which such facts appear, it seems, is a service that philosophical examination offers social and behavioral science.

This is quite an elusive project, and can only proceed if we agree on what the facts are.

The theories emerging from behavioral science do not benefit as much as the natural sciences do from instrumentation, microscopes and observatories. Theorists could proceed more confidently if there were a lens through which one could examine the mind as a moving picture with magnified images that could be played and replayed in slow motion. This might produce consensus as to the pattern and development of the mind's creations, and thus allow discussion to address a common universe of facts.

Some claim that systems of language provide us with such a lens, but if they do, it has not moved scientific discourse of the human psyche very far. The metaphoric properties of the spoken languages may be more bewildering than thinking itself. So it is that the human species creates and isolates a variety of mathematical "languages" designed for the advancement of meaning.

Others have examined organized social systems, the human organizations in which labor is coordinated toward the production of goods and services. That these might resemble human consciousness makes sense. They form through the creative assertion of the intellect, and they are subject to environmental pressures. They consume resources, manage affairs with competitors, and establish priorities. Moreover, they are large, and slow. One job role is apt to engage the capacity of an entire person and they may spend an entire career developing an understanding of what that organization is.

While all human organizations are hierarchically organized, one might expect their hierarchies to bear little resemblance to each other, as each is a specialized adaptation to the products, services and resources with which they are concerned. There was no reason to expect resemblance unless one is particularly receptive to the possibility that external creations, such as institutional bureaucracies, are images influenced by the projective activity of human consciousness. This expectancy led to an important discovery by Elliott Jaques, a trained psychoanalyst.

Jaques began his work approximately sixty years ago. He was, from the beginning, attracted to stratified human institutions. What was more obvious to Jaques than to his predecessors was that a comparison of the hierarchical structures of diverse organizations required an instrument of measure that did not change as one looked from one organization to another. What stood out to Jaques as self-evident in task driven work is that all tasks occupy a period of time, and that the magnitude of an individual's role bore some relationship to the anticipated time associated with the planning and execution of his or her tasks.

With this in mind, Jaques set about to measure the task driven time frame of each individual within a variety of hierarchical organizations. Interviews revealed that people were intuitively comfortable with translating tasks into specific time frames. As expected, supervisors tended to define tasks with larger time-spans than their subordinates. However, a larger comparative time-span was not a sufficient condition for the adoption by one of a managerial role over another. Within the organization, time-span thresholds marked a point of separation between managerial tiers. Manager and subordinate would typically fall within a different range of time-spans.

All of this, though interesting, could hardly be considered surprising. Assuming that a given organizational purpose engaged an environment of a pre-established complexity, one might suppose that certain tasks would have a baseline complexity, that combinations of such tasks would comprise a second layer, and that combinations of combinations would make up another, and so on. What was bewildering about the data, and thus highly important from a scientific standpoint, was that for all the many organizations from which such data was drawn, the time-span thresholds marking one managerial tier from another were the same.

Jaques' investigation naturally led to an investigation of what changes these time-span thresholds represented. They had recorded points of transition between one managerial tier and another at one day, three months, one year, two years, five years, ten years, twenty years, fifty years and onward, to which each was assigned a number, i.e., stratum one (one day to three months), two (three months to one year), and so on.

As Jaques examined the behavior of individuals arranged across these strata, his classification of behaviors along this continuum benefited from a deepening understanding of human organizations, and from the input of several professional orientations, including formal logic and mathematics. A consensus emerged to the effect that the shifts between managerial strata represented qualitatively distinct ways of formulating and processing information, each having a distinct form. The forms were limited in number to four, namely, a "declarative" form that defines objects, a "cumulative" form that gathers objects, a "serial" form that places objects in a sequence, and a "parallel" form that compares object sequences. These, as indicated below, repeated themselves on each of a succession of "orders" of complexity. [2]

The fifth movement in each series has a dual significance. It completes a sequence at a lower order, while commencing a new sequence.

On Freedom: Scientific Freedom

Higher order sequences are thus linked to lower orders. Bureaucratic stratification commences what is referred to as the "third order."

Time-span	Stratum	Processing	Order
One day	I	*Declarative*	*Second/Third*
Three months	II	Cumulative	
One year	III	Serial	
Two years to	IV	Parallel	
Five years	V	*Declarative*	*Third/Fourth*
Ten years	VI	Cumulative	
Twenty years	VII	Serial	
Fifty years	VIII	Parallel	
One hundred years	IX	*Declarative*	*Fourth/Fifth*

Some effort is required inorder for one to acquire a workable grasp of the appearance of these information processing modalities in real human beings. One must learn to count as a juggler as he adds balls to his juggling act. Every addition involves changes of form representing the coordination of a new activity with a repertoire of others. A change of "order" is like a change from apples to animals. Among the objectives of this book is to assist this learning, and in the process, bring much into plain sight which might otherwise have passed notice. We are, in effect, examining the implications of the discovery of a hierarchy in human development. Since it is a matter of describing a processional movement into higher or "better" states of mind, we are examining the discovery of an ethical universe revealed by factual constants in hierarchical organization.

Given a correspondence between the structure of individual consciousness and social organization, a social scientist may, in the evaluation of the "health" of social organisms, imagine that they resemble the health of the individual psyche. Thus Jaques advocates a prescribed order for human collaborative agencies in which there is no perceptible trade-off between the humane treatment of its participants and the efficiency of its productive output.

Jaques offers the terms for such an order to the designers of bureaucratic work environments under the title "requisite organization." Apparently the "mind" is saturated with a bias of sorts in the formation of its social entities. Once such an image is properly formed in this likeness, and individuals are placed within that creation in accord with their capacity, they feel more comfortable, trustful and motivated in their work. The organization prospers. By contrast, transgression against this form is apt to produce distress and paranoia, preventing the organization from delivering the goods or services for which it was intended.

There may thus be a correspondence between the mind and its social creations, which, to the extent it is understood, eliminates much of the trial, error and natural selection associated with the formation of viable social institutions. However, behavioral science, including the area of that science referred to as "economics," has barely started to draw from the well which Jaques and his colleagues have opened.

8. The importance of a discovery of a constancy of form in bureaucratic organizations can hardly be overstated. As we have indicated, such constants otherwise referred to as "facts" are stable points of reference that allow theorists to make more confident claims. This confidence strengthens as corroborating data is developed by other behavioral scientists. Facts—especially trans-cultural constants—allow science to be what it really is—the mental discernment of a reality beneath appearances.

While this discovery did not begin with Jaques' examination of human organization, it undoubtedly strengthened findings and research in developmental psychology already underway. Jean Piaget discovered that children acquire many of the cognitive judgments that adults take for granted over a period of time, and that these acquisitions represent states of improvement. The simple judgment of volume, for example, is unavailable to children until they are able to mentally coordinate the combined spatial notions of length, breadth and width. Before they are able to do so, they tend to identify the tallest container as that which is able to hold the most liquid, even though it may in fact hold only a small fraction of the liquid of a shorter container. As the mental capacity to coordinate judgments increases, qualitative leaps occur in which the child is able to make whole categories of new judgments.

That growth displays discontinuity, i.e., persists in one perspective, and then emerges suddenly into a new one, is completely consistent with the fact that bureaucratic organizations divide consistently into discernible levels. It strongly suggests that social stratification of this nature mirrors the natural progress and maturation of human intelligence.

This discovery of the child's graduated comprehension and appropriation of simple objects points directly to a theory of mind which emphasizes motivated learning, and a world not unlike a trophy case of successful mental conquests. These go unnoticed only because of the value directed at the newest challenge.

We might conclude that phased development presents "improvement" in the sense that growth results in the formation of larger and more comprehensive wholes. This suggests an ethical principle favoring growth and enlargement of perspective—a new version of the philosophy of "bigger is better." A few moments examination of the study of moral development, however, suggest a more nuanced rationale supporting development in general.

Inspired to a large extent by Piaget, Lawrence Kohlberg noticed that in children, moral orientation undergoes qualitative changes as the child matures. Meanwhile, Abraham Maslow, another psychologist, developed a theory of human need which proposes that the satisfaction of needs at a lower order leads to the definition of higher order needs. For both Maslow and Kohlberg, one begins with unmediated self-interest and proceeds to the completion of a socially integrated value. The five stages of this development bear an intuitive resemblance to the sequence of processing modalities within a given order of complexity.

Processing	Moral Stages	Hierarchy of Needs
Declarative (lower order)	Personal Compliance	Physical Existence
Cumulative	Reciprocity	Security
Serial	Altruism	Intimacy
Parallel	Order	Status
Declarative (higher order)	Societal Compliance	Social Existence

Kohlberg's characterization of the first and fifth stage of moral development, are, to use his exact words "reward and punishment" and "social contract." One examines the status of the individual, who compares their conduct with socially prescribed rules, i.e., compliance is sanctioned with rewards and punishments. The other brings principles of social organization into focus, i.e., the compliance of government to principles by which a state can be said to be just or unjust. Maslow uses the terms "physiologic" and "self-actualization" to mark first and fifth strata of need, one which predicates biological existence, while the other defines the achievement of social existence.[3]

There is more than one way to theoretically address a perceived correspondence of this kind. One is to attempt to reduce one to the other. For one who prefers to emphasize the structure of mental events, they may well assert that moral orientations are not more than the reaction of the human organism to a cognitively devised picture of the world. However what if these pictures are themselves fashioned around or in service to an underlying value?

It is similarly difficult to say that values emerging within such a developmental course can sensibly be described as "bigger" and that they appear instead to emerge from something "smaller," i.e., from more unified and integrated response to a challenging social experience. Size may be more a descriptive effect than progenitive cause of "improvement."

Rather than debate over whether human emotive or rational faculties dominate over formation of moral judgments, a more scientific approach might consider whether rational and moral judgments are different aspects of a common experience. That there is no way of recording the interior of a black hole does not prevent an astrophysicist from "knowing" it is there by examining the activity around it. The same problems arise within the human mind. Take the "self" or "identity" for example. Is it a concept or a state of motivation, or neither?

The answer to questions such as these may depend in part on the discernment of a functional ideal or purpose about which diverse mental phenomena aggregate. To put it another way, assuming that emotive and cognitive experiences converge on some sort of "center" or otherwise exist in a state of mutually reinforcing equilibrium, scientific emphasis turns to the value that these experiences serve. A mind thus oriented to this value, presents an image similar to the orbit of planets around a star.

While a notion of "health" is a necessary component of behavioral science, scientists engage the subject reluctantly because it brings them quickly into contentious subjects crowded by speculative philosophers and religionists. The reticence of science about such matters results in theories of behavior which focus less on the contentment of human beings than on how certain kinds of behaviors are adapted to the biological survival of the species. That value, i.e., biological survival, is in many circles the only value that satisfies the scientific requirement of moral neutrality.

Such a requirement eludes the scientist at the very moment he or she expands their universe of possibilities beyond the survival of the biological organism. The survival of one's ego, self and soul would be tainted by moral judgments if science were not otherwise careful to relegate these—as topics of research—to the wilderness of "non-science" until some scientific consensus can be reached as to what these entities are. Even then it would be extremely difficult to scientifically demonstrate that the survival of an "ego" should yield to the survival of a "soul."

Somehow, the moral neutrality of science is preserved where moral judgments are treated as evidence of something else, like the components of material structures of a brain. Although much is gained by mentally rehearsing this neutrality in the evaluation of human behavior, we will at some point have to face the fact that some behaviors are preferred over others, and that choice between such behaviors places science in a difficult position of making ethical judgments. This is something we as scientists find paradoxical, i.e., to eliminate ethical bias while identifying that toward which a human being grows and develops.

9. The prospect that an ethical universe has been discovered or revealed in developmental constants forces one to consider whether and to what extent this universe functions in a state of dependency on matter, or a state of independence sometimes referred to as "spiritual." An attempt to sustain a value neutral pose while laboring to develop a science of consciousness is only a way of putting off the final rendezvous between science and spirituality. Science's avoidance of the importance of religious experience is painfully forced, like two separated in love after a tragic betrayal meeting at a party and talking awkwardly about the weather.

Were a scientist to make a frontal approach to religious experience, it is worth considering what the encounter would or should be like. For some, it is manic, as one having been trained to hold his tongue on religious subjects, speaks of faith as one suddenly released from a rational boundaries. For others it assumes the detached tone of an anthropologist of religious views, speaking with assurance over the deeply held beliefs of others, but avoiding discussion of one's own. For a few more, religious experience is actively renounced as superstition, and is evaluated only to the extent that one might quietly ridicule it.

Despite the tension with which science engages religious experience, it must admit that there is substantial evidence that religious experience is associated with transformative moral experiences. Human testimony relates that addictive attachment to intoxicating substances and behaviors are relieved, broken human relationships are mended, the patient forbearance of suffering is extended, and new identities are discovered and welcomed into nurturing social systems. This occurs, according to report, upon the contemplation and acceptance of a religious ideal offered by many religious organizations.

It may be that such testimony is unreliable and may in many cases be used to manipulate or validate ideological bias. But a scientist would not deny that common to such experiences is a power associated with the contemplation of an ideal, an ideal that becomes more real and convincing when imagined in the life of a "Christ" or a "Buddha."

It would be unscientific to fail to acknowledge the import of much religious experience, i.e., that one's contemplation of religious values, reflected in idealized religious figures, is a potent source of influence. An evaluation of human behavior, even while contemplating the survival of the fittest, must consider the mechanisms by which a human ego is created and maintained. Great religious movements, however badly they distort the teaching of their progenitors, attest to the importance of religious experience in the formation of the human psyche.

Has science, however, failed in some way to properly record religious experiences which otherwise violate scientific precepts of cause and effect? The testimonials of faith go further than the claim of moral transformation, and include changes that challenge the expectancies of medicine, or involve the resolution of ills which are highly improbable. While science in its present state is comfortable with the proposition that one thought may alter another, it will not accept that a change of mind may influence matter unless one can establish a material nexus between one and the other, i.e., brain, neurons, muscles, hands, etc.

There are religious sects in various parts of the world which not only emphasize the importance of miraculous events in the support of their beliefs, but have replaced some or all of medicine with the spiritual treatment of human maladies. The practitioners of this religious discipline, which they refer to as "divine science," are nonchalant in their description of it, and advocate as scientists would, that religious propositions should be proven.[4]

These perspectives serve as a credible basis on which to question—though not necessarily overrule—attempts by science to subordinate mind to matter. Rather, they prompt science to welcome a philosophical attempt to clarify what are meant by "mind" and "matter." Does the material content of a brain in some way serve or adapt to "mind?" Does mind use brain, or is it the other way around? We may not need a final answer to these questions, but it seems that we should, in developing a "science" of mind and its social formations, treat mind respectfully, and allow for the possibility that the relationship of mind to matter may be less one of subordination than of the intersection between one universe and another. This may, in the end, reveal something that systemic philosophers have been emphasizing for quite some time—i.e., a complementarity relationship in which matter and mind are redefined.

One of the difficulties the various scientific communities have with the evidence of religious experience is that it depends on human testimonials, and seems to evade the scientific demand of experimental replication.

There are obstacles to the creation of a satisfying experimental protocol. It is difficult to select among various approaches to prayer those which make an authentic and confident claim to science, i.e., which engage a religious ideal with sufficient completeness that it can effectively dominate consciousness. Related to this is the difficulty of applying prayer to a defined ailment without corrupting the spiritual experience into an instrumentality of material experience. The challenge is to devise an experiment which does not sabotage the morale of effective treatment.

Therein rests the basic paradox of attempting a value neutral appraisal of any religious experience. If the experience requires religious commitment before it can produce measurable consequences, then neutral evaluation of that experience is problematic. The neutral evaluator would have real difficulty in establishing a population to study because he or she is poorly situated to separate real from counterfeit adherents to a given religious discipline. But if he or she fully embraced that experience, the claimed "results" would not have sufficient credibility to survive criticism by a community of non-believing scientists.

If the experimenter is interested in the influence of a religious ideal on one's physical health, as in the "power of love," they would need to remove the experimental subject from a position of control over the experiment. Given the current skepticism of science toward the claims of religious experience, and the difficulty of defining—and thereafter identifying—a suitable population of adherents to a religious discipline willing to act as test subjects, such an experiment is not apt to receive sponsorship from credible authoritative agencies. Even if it were, the experiment might preclude the accomplishment of a reliable negative, i.e., a finding of "no relationship" would likely not convince devout practitioners that the correct religious practices were deployed.

One might, nonetheless, devise a study which demographically compares the health histories of people who immerse themselves regularly in various types of prayer with those who do not. The principal defect of studies which have already done so, in addition to above noted deficiencies in defining similarly situated adherents to a given spiritual discipline, is the failure to exclude other sources of explanation for the weak positive correlations to health that arise from such studies. One might imagine, for example, that religious experience reduces the physical effects of stress and that the resultant better health is therefore physically explained.

Even with these barriers to the scientific examination of the practical efficacy of various spiritual practices, the lack of research in the area is surprising given the volume of documented testimonials of benefits derived from a soulful petition to a transcendental principle. There is, however, enough devotion to religiously inspired value to coax science toward a theory that at least accommodates the influence of a god, or perhaps, the one God, if there were a model in which it makes sense to suppose such influence.

Perhaps the more difficult issue to resolve is whether it is appropriate to refer to theoretical activity which is receptive and accommodating to human spirituality as "science." Does it make any sense to labor on theories of mind or human behavior if those theories are not testable? [5]

10. The answer to this question requires that we address concerns that recur frequently in history. If science is a meaningful advance in human thought, it represents increased tolerance of uncertainties such as these, and an attitude that is humble and patient enough to accept doubt as work progresses toward more satisfying explanations. Science can neither jump to religious conclusions nor exclude human spirituality from its universe. Scientific inquiry may never venture far from metaphysical issues that challenged ancient philosophy, such as whether the rightness of a social universe is linked to the depth of our understanding of it.

If there were somewhere in mind a God directed toward the task of identifying and communicating the right balance of priorities necessary for the development of just and cooperative social systems, it might imagine or create a living example of such a value. Humankind might otherwise find it impossible to think scientifically about itself, in the same way that it would be very difficult, upon receiving a package in the mail, to assemble the various pieces of a product without knowing what it is supposed to do, and having a picture of what it is supposed to look like. Such is the way, perhaps, that religious experience becomes focused on the individual lives of religious figures.

Much attention has been directed toward the investigation of relationships between behavior, neural structures and psychoactive chemicals. It is becoming apparent that injury to the communicative tissues of the brain affects the experience of reality, and that less integrated experiences result in weaker and less authentic senses of "object" or "selfhood." Thus by examining the effect of brain injuries, or chemically based inhibitors to neural transmissions, scientists may clarify the influences necessary to the formation of a whole idea or object in thought.

These useful observations, however, may lead to a precarious theory of mind reduced in quality to a neural mechanism. As argued before, the moral of science evolved from the celebration of an intimacy between the human mind and another supposed "Mind" within which it is. Scientific freedom ascribes value to consciousness, which includes a right understanding of reality. However, a conception of mind that likens it to a assemblage of material structures and chemicals, displaces this vision of unity between reality and human thought, and by sleight of hand, replaces it with a Pantheon. The voices of the gods, the "elohim," are thus redefined as a mind quilted together by matter.

The skepticism of science would therefore become irrelevant in any attempt to understand mind, as it is in any religious system which diversifies deity. Doubting derives from the placement of a scientific value on the uniqueness and cohesiveness of truth which resists a reality rigged with hallucinatory experiences imposed by a material brain. That one wishes to and succeeds in sorting through such effects suggests a mind that values reality. This, in a way, proves scientific freedom.

All this changes as one imagines a "mind" within which human mental phenomena either make or fail to make sense, and that humanity possesses or shares that mind. This attitude humbles the human intellect now challenged to discern "reality." One such challenge is the size and intricacy of the physical cosmos. Another is the historic appearance of individuals who seem to achieve consciousness which goes beyond limitations presumed to afflict the human condition. An evangelist would praise the "mystery" and "omnipotence" of God, and tell the story of creation as they wish it to be. The scientist would sigh at the work yet to do, hoping that some day they may understand. Who is more spiritual?

Some advocate a point of compromise between the evangelist and the scientist. Science, it seems, could be the mental activity that deciphers the intelligent design of God. This point of advocacy bears some resemblance to scientific freedom, but fails. To suppose "intelligent design" is to suppose a God who scribes patterns onto matter. This supposition violates a principle discussed earlier, i.e., scientific freedom does not project human limitation of any sort on God. That would include a God imagined as an intellectual craftsman who makes and unmakes material patterns.

Along with religiously motivated notions of intelligent design go other more pernicious applications of such design to social organization, such as moral fables that use God and punishment to frighten people into conformity with social rules. A God who plays games of creation with a material universe might, as some sort of great director, use the universe as a laboratory in which the human being is something yet to be proven worthy.

The truth is, it is nonsense to say that there is "intelligent" design because "intelligence" as it might appear in God is something human beings are only gradually coming to appreciate. As emphasized before, whether one believes in God is far less important to science than whether one attempts to make God into an extension of human existence. There is nothing at all scientific about that, and results in a tendency previously noted in ancient philosophy—i.e., to impose organization on reality instead of being committed to the discovery of it.

Scientific theories about the human mind progress as confidence in information progresses, particularly that information which suggests a mental world governed by constants. The Ptolemaic view of the world was interrupted by Copernicus, who placed constancy of position in the Sun rather than the Earth. Later the assumption that Space and Time were immovable constants in which physical events occurred was broken by Einstein in the discovery that the speed of light was constant, and that Space and Time were relative coordinates.

Now in the theoretical focus of subatomic physics, mathematically defined symmetries appear to be more real than "particles" one attempts to locate with them. These points of matter that were supposed to constitute an irreducible physical reality do not appear to acquire coordinates of location until an observer asserts themself through a device of measurement. Some theoreticians suppose that all of what is regarded as matter exists only as a field or wave of possibilities until an observer intervenes. This orientation invites consideration of how adjustments to the observatory will affect the reality, and not just the appearance of things.

Thus, to suggest that freedom is constant and that human bondage is superficial error, and may be so proven, has a scientific resonance. If there is such a thing as scientific freedom, then it consists only in one's commitment to adjust its inference of what is "real" to facts, and to demand that any supposition about it be proven. If it appears as though human behavior is attracted to, or gravitates toward something, then scientific freedom would tend to call that something "real" even though it diverged from an imperative of material survival.

As such, this book will regard with seriousness the human capacity to embrace existence for its own sake, and all beings who participate in that existence.

2.

Intellectual Freedom

Everyday language is a part of the human organism and is no less complicated. It is not humanly possible to gather immediately from it what the logic of language is. Language disguises thought. So much so, that from the outward form of the clothing it is impossible to infer the form of the thought beneath it, because the outward form of the clothing is not designed to reveal the form of the body, but for entirely different purposes.

Wittgenstein

1. That aspect of mind that devises plans, tools, and scientific theories is useful to the satisfaction of human need, enough so that we as a species unhesitatingly celebrate it.

As living organisms, we use intellect to search for the things we want, all of which are assembled within a reality that it construes. If we didn't want something, there would be no reason to assemble a reality at all, or even to take notice of it. A concern with where food is located supports the organization of a world into places where food may and may not be found. If one is concerned how the world was made, they might have to organize his universe into scientific theories—places of a different kind.

Without interest, there is no inquiry, and no intellectual endeavor—no intellect. Thus it is, in a way, nonsense to use the term "intellectual freedom" in that the intellect can only exist as a servant of interest. Interest places mind into a relationship with a "world" that may or may not agree with interest. The resistance of experience to interest reveals a world that is independent from desire, and that must be examined and processed by intellect. Interest is the power by which intellect is formed into an instrumentality.

Where consciousness asserts itself in the form of an interest, that interest is an invocation to the intellect to search the world for that thing which satisfies interest. Much, if not most of the time, that thing is found through labor, and occasionally not found at all. The intellect is the best servant of interest when it informs consciousness truthfully, but this does not always occur. The intellect might report falsely about what is there and what is not there.

This is unlikely to occur where one is interested in something basic and primary, such as food, shelter or sex, and quite likely to occur where one is interested in something abstract and intangible such as moral, political or spiritual fulfillment. In those instances where interest not only commands the intellect to search, but also renounces the truthful response of an intellect, the intellect is less than a servant, and becomes the slave of interest. This is the intellect of an "ideologue."

One may also suppose that there is no truth. Such an attitude delights in the appropriation of an academic or professional language in order to engage in formative displays peculiar to the language. Intellect is used to magnify itself. There intelligence loses its essential nature, and proceeds licentiously. Success in this mode is generally welcomed in commercial enterprises that market the proposition that nothing can be proven. This is the intellect of a "sophist."

The ideologue commits intelligence to the rationalization of the presence of a "truth" which is either not there or not provable. The ideologue is trapped by assumptions, until a crisis of some sort causes separation from the interest around which he or she organizes a world view.

The sophist behaves more like an exhibitionist of intellect. Unlike the ideologue, they never experience a crisis of sufficient proportion to break attachment to a given interest, because there is none except that which attaches to the intellect itself. The way out is for the sophist to suffer irreparable intellectual humiliation, and to abandon vanity.

Both phenomena—in their own way—exemplify how behavior we normally associate with "freedom" is corrupted. The ideologue makes the intellect a slave of interest, while the sophist makes interest a slave of the intellect.

The topic of intellectual freedom begins in a sense where scientific freedom arrived, and considers what the mind does once it has achieved confidence in and respect for a substantive reality hidden within or underlying a world of appearances. A shift from scientific to intellectual freedom thus involves a shift from one's orientation to reality toward the description of methods. Intellect is descriptive of the human effort through which mental objects are formed. Its freedom consists in its capacity to find what it is looking for.

Convincing evidence of intellectual freedom, and thus what separates it from its ideological and sophistic contraries, is the growth of the mind toward a more efficient and focused awareness of how to achieve the objectives set before it by human need.

2. In a world divided into places, the successful intellect must have some familiarity with them, and an ability to move among them through

investigation. Since investigation looks for something, an intellect is necessary to set one place apart from another, and to organize those places so that success is achieved. As mentioned above, interest activates investigation.

There have been many impressive attempts to describe and set forth rules of intellect. These attempts occupy a field of knowledge often referred to as "Logic," typically expressed in terms of rules which allow us to distinguish effective and ineffective ways of interacting with the world. At its early stages, logic emphasized consistency in the use of terms, and was offered as a cure to the manipulative use of words to make false arguments seem true.

Along with the scientific revolution, there was a greater emphasis on the formation of mathematical languages, and this led to overt comparisons between cognitive disciplines invented by the ancients for spoken language, and the discipline of scientific thinking. Mathematics and Logic began to appear quite similar, and logic was redefined as a foundational discipline supporting the proliferation of mathematical languages.

The most widely recognized system of logic in current use examines the reasoning processes occurring in language, i.e., declarative statements combining nouns and verbs meant to correspond to a reality of "terms" and "predicates." This logical system builds upon what it calls a "proposition," an assertion about a state of affairs which may or may not agree with reality. Normally we think of agreement between a proposition and reality as "true" and disagreement as "false." The logical system built around propositions is often referred to as "First Order Logic."

Within any given state of affairs there are things we take for granted, i.e., things we already know about and are not interested in investigating, and things about which we are uncertain. Typically, propositions are directed toward uncertainty, though any fact can be the subject of investigation—a proposition. Thus to say "that blue sky," which predicates the sky as an object within one's gaze, amounts to a proposition about it. For that matter, to say "that sky" is a proposition, inasmuch as one predicates the thing above one's head as "sky." The mental commitment in each instance is the same, in that one has taken the trouble to identify a place or object and attach expectancy to it. The expectancy can be correct or incorrect, true or false. The sky may not be blue, or that thing above may not be a sky.

Expectancies apply not only to objects but also to various combinations of them. First Order Logic has established a lexicon around them that upon close examination presents a hierarchical progression resembling the work behaviors evident in the strata of bureaucratic organizations.

It is difficult to understate the importance of this resemblance, because it implies symmetry between the way the mind works on the microscopic and macroscopic scale, or rather, that the mind is doing essentially the same thing in the solution of simple and complex problems. However, since the notational language of First Order Logic was not designed to express this symmetry, the resemblance is likely meaningful only to trained logicians.[6]

Accordingly, it is necessary to devise a language or notation that depicts the stratification of thought in terms that can be easily transferred from simple to complex orders of existence. This cannot be done without contemplating a logic of "objects" that essentially uses the same notation to describe simple and complex objects. Once there is some, although forced, consensus on how we represent objects (both simple and complex) then one can begin to speak meaningfully about how a given domain of objects combine into more complex structures.

The newly devised language may be called a "developmental logic" and makes a number of modest revisions affecting the meaning of logical terms. The most important of these revisions involves the concept of "truth." Because this logic tracks the patterns of thought occurring under pressure toward improvement, this logic may also be called the "logic of work."

There is a difference between the concept of "truth" as a component of a larger theory of logic, and "truth" which is the primary motive and goal of intellectual freedom. In standard logic, "truth" is a quality common to the correct assignment of the value "true" to propositions about the world. Such "truth" is the thing which all true judgments have in common. The truth that is the subject of freedom is something which is sought after, and which tasks and orients one in the course of a sustained inquiry—a truth which defines and gives meaning to personal commitment.

First order logic is concerned with the truth of propositions. A true judgment is one in which one's proposition about the world agrees with the facts revealed through an information gathering activity or investigation. Verification occurs through a test that proposes that if one does something, a consequence will follow. This relationship can vary in complexity.

For a simple proposition such as "the light is green" a simple test such as "open your eyes, and you will see it" may suffice. It is not so easy to test a proposition about subatomic particles, or whether it is better to give than to receive. When Einstein proposed that mass bends space, he argued that evidence of that proposition could be viewed in the event of a solar eclipse, and that the interposition of the moon between the earth and the Sun would allow one to see how the position of the light from certain stars changed in direct relation to the mass of the Sun.

The "truth" as defined in a discussion of "freedom" is, by contrast to true judgments, something over which one is willing to labor. It is embraced by an expectancy of sufficient force that one collects and combines objects together into ordered systems so that one can realize that expectancy more efficiently and frequently. The constancy of that expectancy establishes a baseline by which various improvements along a line of development are measured. Thus to say "the sky is blue" is provably true, but in most instances is not "meaningful" because it is commonly known, taken for granted, and would not task the intellect to organize its environment—and adopt and optimize strategic poses—so that his expectations are satisfied.

When one labors one tends toward improvement upon efficiency. This tending might be attributable to motivation, but it seems to happen without trying, and trying does not seem to make it happen faster. At this moment in a discussion of freedom the controlling assumption is that a more efficient method for obtaining truth is better. Intellectual freedom describes a progression from weaker to stronger ways of finding truth, sometimes referred to as "learning."

In this sense, an abstract description of intellectual freedom may properly be characterized as "truth based" logic, because in this logic truth is a motivating expectancy that activates and justifies one's engagement with his environment. In First Order Logic, motivational elements, though implicitly present, are de-emphasized to the point of disappearance.

3. "Truth" expresses a symbiosis between interest and the object toward which interest is directed. Truth represents agreement between a measured expectancy derived from interest and the measured result produced in one's pursuit of that expectancy. What First Order Logic refers to as a "proposition" is a measured expectancy which can be satisfied—or not—through an act of labor. Logic thus places interest

and object in a relationship where there is a fit between expectancy and outcome. In such a relationship, truth and interest fit together, i.e., have comparable magnitudes. A profound and compelling interest might therefore pursue a profound and compelling truth.

The "truth" that satisfies interest thus depends on interest to obtain its status as "truth." Unless one identifies an interest demanding satisfaction, there is no discovery of truth because there is no basis by which to set one object apart from another. A given set of objects simply are, but without interest there is no basis by which to place "true" as a value on one.

The truth of the intellect represents attainment of the thing which one, at any given stage of development, seeks. If one seeks wealth, then the attainment of that is their truth. If one seeks romantic fulfillment, then the attainment of that is their truth. If one seeks power, then the attainment of that is their truth. One labors for these things because they want them. One seeks it because they either don't have it, or don't have enough of it, which is the reason why it interests them.

Some wonder whether truth is "relative" or "absolute." Intellectual truth, as here defined, relates to the interest that motivates a search. It is "relative" in that sense, but "absolute" in the sense that one is fully committed to finding it, or rather, is constant in his or her expectation of it. At a given moment, intellectual truth reflects the limitation of personal interest, but as an abstract idea, leaves the door open to the contemplation of higher interests, and thus higher truth. If there is an absolute interest, therefore, there is an absolute truth.

The judgment "true" is a value attached to a proposition the test of which brings about the satisfaction of one's interest. So to propose "the sky is blue" when tested by opening one's eyes and looking upward at the sky and seeing blue is "true" because one's interest, "blue," is verified. Common to every such judgment is an act that tests one's expectancy that the satisfaction of their interest will occur. Every such test involves labor, and occupies time.

Labor is generally directed at meaningful or useful interests. Take food for example. The proposition that food is under that rock, up in that tree, or in that bush, is tested by looking in those places. One looks for food, shelter and clothing in an industrialized society by looking for a job, and by doing the job. In most cases, doing the job involves looking for things that interest the employer such that the employer's interest is transferred to the employee. Looking for a wife or a husband is usually more complicated, for not only must one look in places where available candidates may be found, but one may need to find financial security, interests, fitness and a system of values which are suitable to a prospective mate.

If one's interest is in "blue" then truth is really about whether one can find it by looking at the sky. If one's interest is in whether it is likely to rain, then to say the "sky is blue" is more a judgment about the likelihood of rain, which requires among other things, observation of the sky.

This difference may seem trivial when evaluating this kind of proposition, but is quite important in evaluating more complex and important propositions, particularly where there is honest dispute between different people over what they are attempting to determine. Much progress is made in resolving such disputes by simply clarifying the interests giving rise to conflicting propositions, and realizing that these are not proposals about what is, but proposals about how to find or discover a matter of interest.

A common analytical error thus occurs as one becomes overly focused on the verbal content of a given proposition—as in losing a forest among trees. As an argument becomes more intricate, the propositional statements that comprise the argument are subordinated to something larger offered to satisfy the interest which is motivating the argument. Indeed, we cannot determine what it is that someone so engaged really proposes until we deconstruct the argument by way of measured ends and means, and uncover a sensible complementarity between them.

A verbalized proposition—regarded in propositional logic as an essential unit of measure—reveals very little on its own of what the intellect is up to. A proposition illustrates how expectant intention is posed for the sake of investigation and testing, but in most linguistic acts, the speaker is engaged in something quite beyond what a given proposition expresses.

We are apt to appreciate the organization of thought better by treating a proposition as an example of a type of mental occurrence, i.e., an articulated expectancy, and that this occurrence is the basis of observation, i.e., a test by which one determines if that expectancy and reality are in agreement. Accordingly, it is much more interesting to assess the project magnitude with which someone is engaged, which necessarily places our attention on the ability or capacity of an individual to coordinate diverse activities into one task.

The inclusion of the magnitude of judgments as an indispensable coordinate of any intellectual act supports an altered view of the universe in which the intellect functions. Intellectual functioning is sometimes placed within a great abstraction named "reason," but if there is such a thing, "reason" would not be subject to the kind of limitation which marks and defines the contours of a given intellectual act. We might even consider defining "reason" as a limitless continuum within which intellectual limits are expressed.

Thus in the same way that the infinity of space emerges as one's attention shifts from a tree to a forest to an island to a continent to a world to a solar system and so on to a universe, the infinity of reason likewise emerges as one examines the boundaries associated with the finite complexity of a given judgment. In other words, intellectual freedom pertains in part to the establishment of a "space" of sorts called "reason" in which ideas break boundaries associated with a limited intellectual construct.

A perplexing conundrum associated with a presumably boundless or infinite reason or mind into which the intellect grows is that the characteristics that we tend to assign to it are extrapolated. As we listen to others talk about reason and what is reasonable it is probably more revealing about the limitations operative in the way they think than what "reason" actually is.

It is therefore wise to proceed cautiously with claims as to the meaning and content of a transcendent reason. What seems to separate an intellectually free person from one who is not is a willingness to treat their appraisal of the world as an evolving phenomenon. They may arrive at a point they never expected, and begin to see truth differently, quite possibly something that reason discerns, but the intellect does not, or that waits to be known after a certain disengagement from the popular view of what "intelligence" is.[7]

4. Having already noted a symbiosis between interest and action, and a modified conception of "reason" that accommodates that relationship, we can begin to define what it means to establish a limit or boundary on intellect. The sensation of freedom is arguably the strongest when one discusses "crossing" or "passing over" perceived boundaries. This occurrence is driven or enabled by a readiness to construct a reality which drives past those boundaries. This readiness is here referred to as "capacity."

As mentioned in our discussion of scientific freedom, there may be a reality which exists apart from the mind, but our assertion that this is the case is based on the fact that the so called "world" we inhabit fails to conform itself to our wishes. The more obdurate and elusive the world is in responding to interest, the more we ascribe otherness to it, because it is other than what we were led to expect. Conversely, a world in which the interactive variables match expectancy resolves this sense of alienation, and produces atonement, i.e., literally a sense of being "at one" with experience.

There is no apparent guarantee that alienation will resolve well. The world may not have what is wanted, or what is wanted may be there, but too difficult to obtain. Most often, however, taking what is needed arises through the development of proficiency in the management of one's physical and social environment.

A change for the better usually occurs as one learns to coordinate the application of mental operations toward the construct of a more comprehensive view of this "reality." "Comprehension" as a term has special meaning in a logic of intellectual development. It refers to the expansion of one's view while retaining the original focus on detail. It is not a shift of panoramic scale, as in defocusing on the trees to see the forest, but of adding the forest to the trees. Added comprehension derives from an added capacity.

That intellectual activity in some sense reflects an escalating scale of complexity is already strongly implied in contemporary academic textbooks on logic. These texts typically introduce the subject of logic by defining what a "proposition" is, progressing to ways in which "propositions" may be combined, then to the formation of compound arguments from propositions in what are known as "syllogisms," then away from the content of the propositions into the discernment of patterns or "relations" among members of "sets," and finally to the effect of a "universe of discourse" on the perceptibility of those patterns.

These augmentations suggest a hierarchy through which one might progress, provided that points of origin and termination are constant throughout that process. Then if one can correlate a judgment in logic to a baseline in presumed human capacity, it should be possible to correlate an expanding capacity to an expanding complexity of logical judgments. Each stage of development would, under such a formulation, be defined by a limit in the capacity of mind.

The subjective experience of this limitation is the experience of incompatible alternatives. Assume that someone must apply all their effort—their will, strength, and concentration—to apply one mental process to a task with an established complexity, but that there are other mental processes available to apply. Those options cannot be part of the reality of that person's world because there is insufficient capacity to allow those processes to influence it.

To the agent at work, reality presents as a duality, one in which the application of a given mental process is absent, and a world transformed by the application of that process. This duality represents the mental processes that structure experience effortlessly, and a new process applied with effort to integrate with the others. Essential to the growth of capacity is that the practiced application of mental processing skills tends toward the mastery of them. After its mastery, it is assimilated into an old reality, whereupon, the mind tends toward the acquisition of a new processing instrumentality which, because it introduces an innovation, is juxtaposed to the reality we take for granted.

That which is given—which by translation is that which the intellectual mastery of a domain allows us to take for granted—is also at times referred to as "the void." For our purposes, there is no substantive difference in a domain which has no content and one in which mind has no interest. "Content" is an act of selection by a mind which engages what would otherwise stand waiting for us to do something with it. The human intellect is a content generating instrumentality.

In the course of a given life, this experience recurs in a number of familiar and significant intellectual acquisitions, as in learning to use sensory organs, to walk, to speak, to play a musical instrument, to develop a trade, and manage a complex organization. These acquisitions suggest a perpetual duality between a world at rest—an object which we construct without having to try—and what we bring to it as trying. The fact that this *trying* is hit and miss until we master it is the experience from which an external reality is constructed.

This duality, it seems, resonates at a more basic level than the Cartesian dualism which dominates enlightenment philosophy. That division, which separates the thinking "I" from an exterior reality, has always been an elusive—though useful—construct. The problem with it is that externality is an inference derived from the resistance we encounter within experience. We suppose it to exist because our experience is not exactly the way we want it to be.

The same reasoning is applicable to other dualities we have come to regard as fundamental, such as a duality between "false" and "true." To standard logic, the judgments "true" and "false" are qualitative contraries, i.e., "not-true" equals "false" and "not false" equals "true." "False" and "true" have equal value, and appear depending upon the application of a mental act called "negation" as two sides of the same coin. But what substance would such a duality have if that "space" could only accommodate one notion at a time, i.e., true or false? The judgment of "true" could not hold or remember "false," just as "false" would not hold or remember "true."

If truth was a form of light, and one entered a dark room, their expectancy of light would suffer disappointment, and would therefore assign "false" to the room. By contrast, upon entering an illuminated room, one's interest is satisfied, and they assign "true" to the experience. At this simple level, there is nothing that can alter the satisfaction accompanying a light room and the disappointment accompanying a dark one unless one shifts their interest from light to dark—e.g., from a desire to see things to a desire to be free of visual distractions. If one's capacity to expect something is exhausted in a single moment, that shift is untenable.

The usefulness of the judgment "false" consists in the benefit we derive from the recognition that "true" did not occur. "False" is a reason to reject—and not to repeat—an expensive activity, i.e., an activity occupying capacity, effort and time. However, where the magnitude of capacity is spent on testing one's expectancy, "false" has no formal status other than to signal error. The value of "false" derives from and is secondary to the primary value we associated with a search for truth. There is no interest in the absence of truth except to avoid failed attempts to find it.

Although engagement of the void begins through a mental assertion which carves out a position within it, we must consider how one moves from the occupation of one position to many. Is the world like a balloon which is filled with such inventions? If not, then we might suppose that such assertions follow a pattern through which whole objects become parts of larger wholes, and so on, until an ordered universe appears.

A progression of that nature would, nonetheless, take heed of how every increment of magnitude moves one through such a progression. At each stage we ask the same questions we asked of "true" and "false." How much available capacity is there, how are limits experienced, and how does the introduction of new capacity traverse those limits?

5. If there are qualitative improvements in human development, our sense of what divides the world—i.e., what we construe with and without effort—will change from time to time. Research in human development strongly implies a limited set of logical operations which integrate into the formation of an "object" and that these operations integrate into consciousness for a given type of object individually and sequentially. Thus if one is preoccupied with very large and complex

objects of a given type, a process of integration is apt to occur over a long period of time as one learns to apply mental operations to the comprehension of them.

As we have suggested before, there is much research that this organization is a phased process, and that these phases have been observed in childhood development and in bureaucratic structures as modes of information processing referenced as "declarative," "cumulative," "serial," and "parallel."

Standing alone, these descriptive classifications do not provide a very satisfying account of this emergence. They do not convey any sense of what the "its" are which are being processed, i.e., the entities which are being declared, accumulated, serialized and paralleled. Additionally, these descriptive terms fail to identify what the mental act is, the introduction or deletion of which produces a change of the form of information processing. They only weakly communicate how various stages of development reflect an expanded capability, and thus represent intellectual progress.

They do not, in other words, amount to a *developmental logic*. What can properly be termed "logic" of intellectual development communicates the necessity and the value of the movement. In the logic of propositions, the necessity of combinative mental processes is shown through a calculus of propositions, but this calculus does not directly describe how intellectual development occurs.

Here is a calculus which addresses that need to track the improvement of intellectual functioning.

The notion of "improvement" requires a baseline in what has been referred to as "capacity." Capacity identifies one's ability to perform a task. By accepting any given capacity as a baseline, one theoretically takes the task or act that requires all of that person's capacity as the "size" of the baseline task. Then as one's capacity increases, one should be able to perform more complex tasks, while old tasks which seemed complex appear simple.

Take as the starting point in a logic of development a reward for labor, i.e., the thing in which one takes interest, together with a task having a baseline magnitude. The baseline is defined as a capacity of 1, which corresponds to a task having singular magnitude. This may also be regarded a "point" or "presence" within a void. Thus a point of existence—experience not yet knowing that this is what it is—is represented within a void as a singularity ".". [8]

There are a number of ways of comprehending that presence which is here referred to as a simple object. The first of these occurs at a capacity of 1 through the introduction of interest. The experience of interest, as previously discussed, is fundamentally different than the experience of object. An object standing alone has no value. Value derives from needing or wanting something from experience. Value thereby converts a neutral experience of object into a positive experience, one of agreement between interest and object—Truth.

Interest thus requires that we introduce two new signs into the logic, one for interest "( )" and another representing the transformative effect of interest on an object when that object satisfies interest "+". The experience of "true" may thus be signified as "(+)"—a state of agreement which alters the status of "·" to "+".

The statement "the light is green" is true if one performs the task of opening one's eyes, looking at the light, and seeing that it is green. "Green" is ( ), the matter of interest realized in the performance of the act (+). The statement is false if one performs the act and their experience fails to validate or support the thing of interest. False "·" is thus a state of disagreement with or exclusion from interest.

The same signs might apply to more complex engagements, with appropriate attention to maintain consistency between the size of the interest and the size of the activity. If for example one is interested in the manufacture of lumber, then the task may well involve locating an appropriate tree, wielding an axe against it, carrying it to a mill, and selling it. It would still have the same symbolic representation of (+) a program which stood apart from or was distinguished among others as successful.

Value is a state of potentiality ( ) in which an object either agrees (+) or fails to agree "·" with interest. Thus a rule of exclusion, i.e., a thing cannot both exist and not exist, which many logicians regard as irreducible, actually derives from a rule of procedure, i.e., expectancy precedes existence. The resultant binaries—*existence* against *nonexistence* or *true* against *false*—reflect the intervention of a value which allows one to set *something* apart from *nothing*.[9]

Interest therefore impresses value upon the void as a *logos* or ruling principle which supports practical engagement with the world. The point is to set apart one or more objects from others through selection. The binary character of objects does not inhere in the objects, but is the result of value driven choice.

6. As simple as this idea is, its application to first order logic can be elusive. While the interest that supports a developmental process is an emotive state, the context of action, and the limits of capacity produce formulated objectives—an intention framed within the confines of a task. This frame of reference is sometimes referred to as a "plan," which indicates intention without providing a reason for that intention.

One can often, through considerable effort, discern the motivation that governs a complex pattern of behavior, but only partially from what another chooses to say about what they are doing. We are not that good at explaining ourselves, and avoid the conscious acknowledgment of what motivates us. Even upon understanding ourselves we often find it necessary to conceal the desire motivating the plan from the view of others. Language is thus not a very facile instrumentality for observation into the logic of work.

While the dissection of language leads to many valuable insights, it reveals little about what someone is doing, and how that differs with what someone else who functions within a different set of skills is doing. The logic of work—i.e., action pushed toward limits—reveals a correspondence between individual capacity and the potential constructs resulting from that work within a field that surrounds that individual.

The field consists of "places" that represent performance options available to a baseline capacity of 1. In an environment or field in which all but a few—or one—of these options holds or agrees with one's interest we can see that (1) the model engages one to work and (2) it is unclear whether one will be able to satisfy their interest unless they can come up with a way to select the right option. The model thus adopts at its core the primitive biological metaphor of an organism in search of what it needs.

It is also a model of what in mathematics is called "information processing" in as much as a field of objects is experienced by the intellect as "information" and that the "processing" of information is descriptive of the intervention of interest upon that field. Much depends upon how the information is organized in the searcher's mind. At a capacity of 1 the world of objects presents itself as follows:[10]

$$1(\ ) \qquad\qquad\qquad ......+.$$

If the objects were arranged vertically (as a composite whole) then we might imagine that the satisfaction of interest would be immediately visible, and thus attainable in one act. Or if the search could subdivide the field, fewer acts would be needed to find +.

1 2 3
..../../+/·

But we ask the same question we have before, where is the capacity to organize the field in ways such as this? At a capacity of 1, we see + but the actor does not. Since each object requires one's complete capacity to discern, they discern but one. We haven't yet allowed the capacity through which such organization takes place. Under present limitations, there is barely enough capacity to process the information represented by a single place. As a consequence, the search among these places is random, and could last indefinitely, or as long as error is repeated.

This beginning stage defines what we have previously referred to as "declarative" processing, i.e., work behavior preoccupied with and limited to singular constructs. A side-by-side arrangement of terms is one way, at least, to summarize and explain a correspondence between standard and developmental logic.

First Order Logic (first level)	Developmental Logic Capacity 1
Proposition	Interest ( )
Term	· Object (Absence)
Predicate (Proposition)	( ) Interest (Presence: primary interest)
Negation (Neutral to True)	· to (+) Distinction
Excluded middle	Presence cancels absence

We have discussed a primitive divide between what is given and what mind introduces to what is given. Hence the logical definition of "term" and "predicate" compare favorably to "object" and "interest" above. A *proposition* is the verbal effect of interest, an interruption upon what is taken for granted with an imagined state of affairs—imagined because of interest.

While *negation* is simple to apply, it is nonetheless difficult to state with clarity where it comes from and what it really means. Here we recommend that it is descriptive of an intervention which alters a neutral state of affairs. Thus to say "not-false" is the same as to change absence to presence through the assertion of value where none previously existed—a value which allows us to set one "·" from another as "+".

The fact that "negation" suggests an anti-positive and disagreeable orientation is unfortunate, given that negation as *distinction* occurs through the affirmation of value. However, the problem goes well beyond a misleading terminology. Negation fails to identify the direction of intellect, i.e., that *true* represents—and cannot occur without—an affirmation of value. At the first level, thought moves from false to true. This allows us to assign a numeral "1" to the event, because we have introduced one and only one intervention.

The reverse movement would not be possible in a logic of work because a reverse expectancy is not yet available. False is not recorded or remembered, and is replaced by preoccupation with another. The negation of such failure does no more in mind than release it back to what it was—there is no assignment of "false."

The "excluded middle" is a rule that prohibits simultaneous assertion of a fact and its negation—i.e., "A" and "not-A" may not coexist. Presence simply is and doesn't need a rule, but the fact that "presence" exists in a relative state of comparison or agreement, presence can exist as a matter of degree. The logic of work thus avoids conceptual dualism a number of logicians offer as a criticism of "Aristotelian" systems which—according to their argument—force choice between white and black, light and dark, etc.[11]

That interest alters the status of objects and of fields of objects allows us to avoid reflexive paradoxes which are a problem in First Order Logic—and which we will more completely address below. Basing logic on the essential difference between interest and object, and the impact of one upon the other allows us to see logic as descriptive of a broader spectrum of human organizational constructs. One example is certain quantum phenomena, wherein particulates organize instantaneously upon the introduction of observation and measurement.

7. Whenever an improvement is introduced into a logical system, that improvement may be translated into a duality—i.e., changed versus unchanged, active versus passive, affirmative versus neutral. It is, in a sense, misleading to refer to these as "dualities" because the duality is created only through the insertion of a single influence which methodologically alters the logic.

We have not introduced two new ideas, but have through the introduction of one new activity, marked a divide of sorts between a world which includes that activity and one which does not.

A capacity of 1 functions under a very restricted sense of time, as a monocular capacity, limits the formation of any but a single moment. Past and future translate easily into that which has happened and that which has not happened, a distinction which occurs only through the mental insertion of recollection.

This is readily apparent in the shift to an assumed capacity of 2. If truth begins with a capacity of 1, then consider what change occurs when capacity increases. The logic of discernment is both enabled and bound by a capacity of 1, a fact which one is incapable of appreciating until the logic assumes that one may effectively apply a second mental activity to objects on a scale of complexity defined by a capacity of 1. The new point of arrival recollects those objects, having acquired correct assessments of their value—or lack thereof—in an assemblage of ideas that is also referred to as "knowledge."

As in the case at the first stage of development, the active intervention of recollection occurs by trying. It is thus supported by an interest in doing so, which is to say, the mind wants to remember its errors in order to avoid the repetition of them. However, the strength of this motivation is, at this developmental event, an interest derived from one's engagement in finding +. Remembering is not the truth one seeks, but an instrumentality in the more efficient realization of truth.

In a logic or calculus of development, improvement is measured in relation to an interest that remains constant. In that way, the utilization of a mental processing skill is regarded well because it brings about what is needed more efficiently. The application of new skills is desired because one holds to an interest which confers value on them. The constant interest which engages the world may be referred to as a "primary" interest while the desiring which accompanies the application of a new information processing mechanism may be referred to as "secondary."

The secondary value introduced to the search for truth is "knowledge." Along with the intervention of this interest, the world of possibilities divides into two separate collections, one for which the status is known and another in which it is not known. The new act can be drawn with brackets [ ].

$$2[(\cdot)] \quad \cdots + \cdots \quad \text{proceeding to} \quad 2[\cdots(+)] \cdot$$

Here one sees conscious assignment of value to a distinction—[(·)] is known to be false and [(+)] is known to be true.

In the science of organization, "cumulative" describes the mental processing in the second organizational tier. In First Order Logic, the most primitive way to group objects—also known as "propositional variables"—is to gather them within what are known as "propositional connectives." One of these is called a "disjunctive" ("or") connection. The other is called a "conjunctive" ("and") connection. Two others seem to derive from the application of "negation" to these connectives—the "conditional" ("if-then") connector and the "bi-conditional" (if and only if) connector.

These connectives may be arranged and compared beside the transformative sequence of a second phase of logical development.

First Order Logic (Second Level) Connectives	Developmental Logic (Capacity 2) Knowledge [ ]
Disjunctive (or)	(·)+ non-retained
Conjunctive (and)	[ ] retention (secondary interest)
Conditional (if/then)	(·)+ [·(+)] Analysis
Bi-conditional (if and only if)	Knowledge cancels Ignorance

Propositional connectives have truth value, subject to the truth value of their constituents. We can say that the expression "p and q" is true if we are right in what we propose about all of the constituents of that expression. We can say that the expression "p or q" is true if we are right in what we propose about just one. These two expressions thus present a duality—one requiring correct proposals about all the constituents and the other not.

Moreover, a conjunctive expression allows one to simultaneously apply "true" and "false" to a given assertion thereby overcoming the rule of exclusion encountered at the first level—"A" or "not-A," but not both. Take a group of rejected utterances (pqrs) combined into a conjunctive expression: "not-p and not-q and not-r and not-s". In each case the expectancy has been modified by adding a negative expectancy. Two interests combine in making the judgment "not-p": P's affirmative expectancy or predicate (a primary interest) and a negative expectancy "not" (a secondary interest).

The utility of these connectives in First Order Logic consists in the fact that human beings express themselves by combining various terms. That Logic uses the spoken language as the source for the derivation of primitive ideas—as that to which symbolic languages must adapt—may tend to conceal rather than reveal their derivation. The examination of the second column, however, suggests a more compelling rationale for the application of a "conjunctive" to experience. It represents a distinct problem solving mode that has efficiency suitable for comparison to behavior representing the absence of it.

There is a nearly universal tendency in academic textbooks on logic to assemble conditional and bi-conditional operators together with disjunctive and conjunctive operators. From the platform offered in First Order Logic it is not clear why this is so. But the right column representing the logic of work strongly suggests that a conditional operator represents the transformative event—from non-retained to retained—in a fashion similar to the transformative act appearing in negation at first level logic—from absence to presence.

Conditional operators are in fact equivalent to a partially negated or modified conjunctive operator. The expression "if p then q" is the equivalent of "not-p or q" which is the same as "p" distinguished in "pq"—"(p)q." Such equivalence is proven through the use of truth tables arranged as follows:

p/q	not-p or q	if p then q
t t	t	t
t f	f	f
f t	t	t
f f	t	t

The replicated pattern (tftt) of the second and third columns is an accepted proof of the identity of the two expressions. Willard Quine—a well regarded philosopher of logic—argued in his Introduction to Logic that it is error to logically confuse the conditional expression "if p then q" with the implicative bond "p implies q." The conditional expression means only that q follows after p.

It is equally significant that a "bi-conditional" (p if and only if q) states a rule of exclusion between the disjunctive and conjunctive expression. The defining requirement of a bi-conditional expression is that all within the grouping share the same truth value, i.e., true for one includes true for the other and false for one includes false for the other. As in the rule of procedure identified at 1, this expression excludes a middle position. Accordingly, the bi-conditional intuitively conveys a state of completeness requiring the assimilation of all unknowns. That the exclusion (one but not both) is equivalent to a bi-conditional is proven in the last two columns.

| | A | B | | | Not | A or B and | |
p/q	-p or q	-p and q	A or B	A and B	A and B	not A and B	p if and only if q
t t	t	f	t	f	t	t	t
f t	f	f	f	f	t	f	f
t f	f	f	t	f	t	f	f
f f	t	f	t	f	t	t	t

As we observed at **1**, a rule of exclusion is not necessary to the logic of work because existence does not need a rule to exist. One records its appearance, and takes note of the change that occurs. Since there is only one thing new happening—the integration of motivated recollection—here again we may assign a numeral **2** to this occurrence.

8. Though the movement from declarative to cumulative thinking is a liberating event, the liberation includes new confinement reflected in a seemingly impenetrable dualism, i.e., known or unknown, but not both. We know intuitively, however, that this restriction is superficial. Much in thought is not known, but comprehended nonetheless because it is implied in what we know. It isn't necessary to take the time to check on it. If we had to do so, we would find it quite difficult to manage our affairs.

Since we had suggested that mental progress consisted of combining parts into larger wholes, it might have been appropriate to view a progression from 1 to 2 as a simple matter of combining one object at 1 "." into two objects at 2 ".." But engaging in a union of this sort is unauthorized within a logic of work until we assert or articulate the capacity to do so. One can hardly talk about implicative combinations among objects until one is first comfortable with forming in thought at 1 and recording their diversity at 2.

At 3 boundaries are drawn with the addition and integration of another activity, and the world as it once appeared at 2 is not the same. The third activity is one of inference or implication. One can know what a task will yield without performing it only if they know that the yield of one task implies, without performance, the yield of another. This occurs as one appreciates that two different tasks are not entirely separate, but may be linked together. One is known and the other, though unknown, is inferred.

Such knowledge, however, requires the performance and evaluation of at least one part of a linked pair of tasks, and, in addition, the attachment of the other to it. Such is the way one, for example, will know by seeing that the light is green, that the light on the intersecting street is red, or by seeing a group of leaf covered branches on a tree that a trunk which connects it to the ground lies somewhere underneath. This is a matter of importance to a parachutist who knows better than to steer into a mass of leaves.

This change of orientation is noticeable enough to others to support a promotion, for example, into different bureaucratic level. In a law firm an individual generally referred to as an "associate" knows how to research the law. But part of becoming a "partner" may include anticipating how a court will rule when the research is challenged. While this ability may arise as new in a setting involving complex tasks, such as a bureaucracy, it is likely to have appeared before in less complex settings, such as the surprise of an infant confronted by an adult in a game of "peek-a-boo" or in the toddler who struggles with apprehension over what lies under the bed or in the closet.

In order to examine this change, it is necessary to introduce a third sign "→" to signify the connectedness between two operations. The expression ·→· thus signifies that · includes ·, and the expression (·)→· signifies that the performance of ·, (·), resulting in a judgment, likewise includes a judgment of the other not performed. Now there is search among possibilities representing improved efficiency.

3[(·)→·]   ·····+·   proceeding to   [·→··→··→·(+)]·

The improvement is manifest in the reduction of acts (·) necessary to realize Truth. The simplification of the field from a field of many ······+·, to a field of many bound within a few occurs because one is capable of inferring unperformed acts. They have gone a stage beyond the simple act of remembering.

There are limitations to this improvement. At this level, one applies a newly discovered ability to reach beyond what they know. This does not involve the application of a "theory" of connection to "other" but is limited to the perception that "other" is connected. This is the experience of "insight," a form of apprehending connection without articulating a reason. It goes in only one direction, i.e. from something known to something unknown, but inferred. After one reseats him or herself, they do it again. Thus an "unknown" is either inferred by its connection to what is known [(·)→·] or it is not [(·)]·, but not both at once. A composite judgment cannot be made without connecting to what is already known.

Thus [(·)→·] is permitted, but [(··)], [(·)·] and [(·)·→] are not. In each of the three latter proposed judgments, one associates two operations or ideas ·· with bare supposition when there is no connecting link →

which includes something which is already known. Thus the phrase [(·)]→·····+· is permitted because one does not suppose that they know something other than [(·)] while + remains buried within remaining operations. It is better to refrain from extending a claim of knowledge [] any further than (·) unless one can identify one and only one "other" to which it connects.

That is precisely what is being addressed for the first time at a capacity of 3, the limiting rule of which is that one can only knowledgeably infer from what they know. A single act of knowing [(·)] thus serves as a platform or axis from which a knowledgeable inference of inclusion [(·)→·] is drawn. In order to include an unknown to what is known, one must do something in addition to what they did before, a thing denoted by the sign "→".

In doing so, one engages in the logic of demonstration, something that may also be referred to as "serial processing" because of the anticipative—but mono-linear—character of the work. It is mono-linear because the native limitation of 3 allows but one platform of extension. Within the lexicon of First Order Logic, following a definition of connectives, one learns to prove unknowns.

First Order Logic (Third Level)	Developmental Logic (Capacity 3)
<u>Inferential calculus</u>	<u>Insight</u> →
Prime formula (variable)	[(·)]+ Supposition
Composite formula (function)	→ Implication (tertiary interest)
Extension (Modus Ponens)	[(·)]+ to [(·)→+] Demonstration
Principle of induction (Validity)	Implication cancels supposition

The phrase "propositional calculus" is meant to describe the derivation of a logical expression from another expression or assemblage of expressions.

In a "calculus" the content of a proposition is treated as an irreducible object, and referenced with a symbol as a "formula" or "variable." A prime formula is attached to another through a transformative mechanism called a "function" or "operator." In the calculus of "syllogisms" we hear of "minor" and "major" premises leading to conclusions—which are meant to distinguish a state of affairs taken as given or known from one that connects that state of affairs to another that is not.

"Modus ponens"—often called the "rule" of "inference" or "extension"—sets forth the most simplified expression or description of what occurs when one fact is inferred from another. It consists in the statement "p, and if p then q, therefore q"—a conclusion drawn from the attachment of an implicative premise to a declarative one. While it is often called the "rule of inference" the "rule" it represents merely depicts the integration or addition of an implicative premise with one that is known, such that knowing one includes another.

The "rule" does not consist in the act of describing this integration, but in clarifying the separation of one state of existence from another. The column on the right hardly needs to express this occurrence in terms of rules, because the depiction of the activity communicates the transformation without any need to remind the observer that the introduction of a process has altered the form of the logic. In a logic of propositions, however, the unwieldy character of common language adopts the term "validity" to emphasize the need to mentally separate what is known from what one attaches to it.

"Validity" is a term used in almost all logical systems to describe the standard by which an argument is evaluated. A "valid" argument is one in which the conclusion must be true if the constituent premises are true. To say that an argument is "invalid" in First Order Logic is to say that the rule of exclusion noted above is violated. An "invalid" argument is a claim to know an unknown without linking to what is known.

One thus commits a logical fallacy by failing to maintain procedural separation between the thing which is given and that which connects to it. Consider the following: "1. John is a person (x); 2. if John is not selfish he is a saint ~(y)→z, 3. if John is a person he is selfish (x) →y, 4. John is a saint (z); 5. therefore John is not a person, ~(x); therefore John is both a person and not a person (x) and ~(x)." The problem with this proof is that "selfish," is used differently in cases 2 and 3. The link between "y" in 2 and "y" in 3 does not exist.

The fallacy represented here—sometimes called the "fallacy of equivocation"—expresses a need to exercise caution with semantic ambiguity posed by common language. An implicative premise either links to a declarative premise or it does not, and one preoccupied with the validity of an argument is apt to scrutinize it in order to confirm that the imperfections of spoken language have not in some way corrupted it.

It is, nonetheless, useful to have a simplified evaluative template available when addressing what may appear to be complex positions and disputes among the proponents of complex positions. Validity allows us to assess whether someone has stated an argument, i.e., that certain conclusions follow from the establishment of certain facts. Once passing the test of validity, we shift to the premises, looking first at what we know to be the case, then to the principles that connect those facts to others. Disputes generally center on the parts of the argument that require inference from a known to an unknown.

9. The real improvement in the change from 2 to 3 was a change in the comprehension of other—i.e., from an undifferentiated association of objects not yet discovered to the identification of a single other implied by the comprehension of one. The singularity of an anticipated other is an imaginative construct which allows one to avoid direct engagement with that "other" and thus move through a field of possibilities more efficiently.

Yet as the sense of struggle over the identification and comprehension of other eases, we may become interested in eliminating the sense of "movement" altogether and find the reward immediately. This is not achievable without the introduction of another mental activity that again changes one's method of information processing. The question we therefore ask is: how does the mind immediately comprehend objects which are not connected directly to a single "here"? How does thought jump ahead with confidence toward the imaginative discernment of an object "located" within a "system" of objects?

At 3 the mind is not capable yet of seeing the object with which it is engaged (·) as having a relative coordinate which locates it among others. "Location" at 3 is not a meaningful idea except as a way of distinguishing "here" from "there." Since here (·) is singular, there is a matter of the direction to which a given platform extends, i.e., ·← (·) or (·)→+.

The mental process at 3 is described by practitioners of Requisite Organization as a "serial" process because of the mind's inability to disengage from its position in series. Anticipation is unidirectional because there are only two locations which mind embraces at a given moment—"here" and "there." "Here" is where one is and "there" is where one extends, their "direction." Moreover, if the place called "there" is a few events away, it is reached or realized only by way of passage through intervening steps.

The phrase "parallel process" was meant to describe an improvement over lineal progression of this nature by emphasizing a state of mind that is able to organize many lines or series into a single pattern. Within an ordered pattern a point may trace numerous pathways, but these pathways are governed by the pattern. One's awareness of the patterned whole may be called one's "understanding" of all its lines of connection.

While one can suppose that a field of possibilities ·····+· is interconnected, they cannot, from the platform of what they know at a particular moment, achieve truth without an added concept which would permit them to pinpoint truth from among many potentialities. The expression (·)→·····+·, while suggesting that truth is connected to what is known, fails to indicate where it is, and thus requires that one parse the field until its location is discovered. The order of the field was not perceived, and without such perception, the location of one among many cannot be identified.

An argument, i.e., a platform from which an insight is reached, may now be thought of as a "stand" [(·)→·]. The field then becomes a collection of implicative relationships ·←·←·→·→· and so on. Now consider a sign →← opposed to ←→ which denotes the order of diverse moments in a field such that there is one and only one place that each can occupy. Order →← denotes the unification of many under one. Thus if [·←(·)→·] denotes a pair of interconnected stands— ·← (·) and (·)→· —then ←(→···←)→ denotes an "understand." A capacity of 4 thus defines "understanding."

Important to this notation is that the "understand" →← fits within or "under" the signification of the distinction ( ). This is because the unifying feature of an understand sizes it to fit within the extension of a distinction. The distinction fits a singularity, which is what the introduction of a relation of order does to a given manifold of objects. There are two features of major importance to this added act. There is first a noticeably more efficient search within the field.

$$[← (→···←)→ ←(→···←)→] (+)$$

This yields + in three acts. Second, there is progression toward order which understands the entire field, or rather,

$$·←(→···←)→·to ·← (→·····←)→+ and to ←(→·····+·←)→$$

which is to say that once it appears that a unity may be construed out of any number of possibilities, there is no limit to growth except the numerical limit in the field of objects. Such is the logic of "theory."

Over a period of time, the field is organized such that achievement of truth is discovered through the performance of just one act. Something can be both implied and not implied through an act of reduction, or rather, the organization of several objects into a single pattern. They are not implied "from" a single platform but are included as part of a pattern.

This is a familiar event. Much, if not all, of what one takes for granted in the world involves labor of this sort. A tree, a recurring example, was once only a collage of sensory data, out of which roots, a trunk, limbs and leaves appeared. It is later discovered that roots and trunk are connected, as are trunk and limbs and limbs and leaves, and that, in fact every part seems to be connected with its adjacent part. Yet it is still just a field of connected spaces until one comes to the realization that underlying its connectedness is a pattern.

Understanding is thus not more than the discovery of pattern within a field of possibilities. Again using the geometric metaphor of a line, there is a transformation from an unordered relation

$$1.\_2. \quad \text{to} \quad \begin{array}{c} 1.\_2. \\ \backslash \ / \\ 3. \end{array}$$

which introduces a triangle where there was once a formless line segment. Since this is not an end to the process, larger more encompassing forms may take over. Thus is a continuation.

$$\triangle \text{ to } \square \text{ to } \bigtriangleup \text{ etc.}$$

One can use a geometric metaphor to appreciate that in any number of contexts, form is, by definition, the discernment of a unifying pattern within a field of objects. One may thus think of a movement from →1·2·3·← to →1·2·3·4·← to →1·2·3·4·5·← as equivalent to →△← to →□← to →⌂←, each vertex (corner) having an assigned position in space, can immediately appreciate that "orders" of this nature might be viewed as foundational to mathematical activities exhibited in numerical and geometric extension.

Complicating this metaphor is the fact that different fields are more or less susceptible to the arbitrariness in the choice or imposition of order. A physical object is, to a degree, what it is, and its pattern is a simple matter of discovery. A bureaucracy, a collection of persons committed to various tasks may, to a degree, be told what the order of their tasks will be.

Upon the discovery that there is a structure to arguments, one is therefore apt to become interested in the way more complex arguments are arranged. This marks a point of transition between what many refer to as "logic" and "mathematics." The latter of these entities is preoccupied with the development of languages for the discernment of patterns. To this end, a logic of "relations" has evolved which may be compared schematically with development at a capacity of 4.

First Order Logic (Fourth Level)	Developmental Logic (Capacity 4)
<u>Binary Relations (R=relation)</u>	<u>Understanding</u> →←
Symmetric	[·←(·)→+] diversified others
xRy or xRy	
Anti-symmetric	→←order (quaternary interest)
xRy and not yRx	
Transitive	[·←(·)→+]then[(→··+←)]Theory
If xRy and yRz then xRz	
Reflexive	order cancels disorder
All xRx	

The assembly of terms used in propositional or "truth functional" logic at successive levels has benefited significantly from the work of R.O. Gibson and D.J. Isaac in an article published in 1978. It must be observed, however, that the use of "truth functional" logic to describe the change which occurs is awkward—defining the mathematics of binary relations in terms of "columns" of truth functional judgments:

> Whereas at level (3) the emphasis was on extension as relative, as successive columns, at level (4) we are concerned with relationships between columns seen as aspects of principle. Thus at level (4) columns are not considered piece-meal but as ordered pairs in combination.[12]

One can hardly ignore the fact that the mathematics of binary relations—regarded as part of set theory—emphasizes a reduction to four types of relations and that these have some resemblance to four connectives identified at a second level of First Order Logic (disjunctive, conjunctive, conditional, and bi-conditional). This quadri-partite assembly thus suggests a continuity of form between levels—i.e., that something similar occurs at each level. This continuity is easier to comprehend—and thus probably more aligned with human experience—by thinking of logic in terms of the management of information (developmental logic).

Level Two	**Level Four**	
Disjunctive	Symmetric	Given state
Conjunctive	Anti-symmetric	Improved state
Conditional	Transitive	Integration
Bi-conditional	Reflexive	Rule of Completeness
Capacity 2	**Capacity 4**	
Unknown	Un-ordered	Given state
Known	Ordered	Improved state
Analysis	Theory	Integration
Knowledge	Understanding	Completeness

As we have observed, the introduction of a new mental operation by its nature establishes a duality between (1) pre-integrated and (2) integrated states of affairs. That duality accounts for two expressions, while two others are descriptive of (3) the process of integration and (4) a principle of completeness which allows only one of (1) and (2).

Thus a neutral state reflecting a diversification of "other" stands in contrast to the revision present in their formation into a single *order*. In the area of mathematical relations "symmetry" and "anti-symmetry" respectively signify the absence and presence of order.

Symmetry is descriptive of systems where there is no importance attached to the positions which extend from a common origin or axis. Given a common axis (·) one's extension to →· or ·← has order only in the subjective experience of recall, i.e., doing one before the other. The significance of such objects is only to exist *there* as opposed to *here* and hence one ascribes equal and interchangeable status to them. In the application of that relation, there is no meaningful difference between them except the fact that they are partitioned.

A change occurs, however, when these points of extension are brought within a common space, i.e., resting within the space defined by expectancy (·) instead of resting beside it. To recall, space is a metaphor, i.e., an interest in which a task, object or location may occur. A system of such things thus do not appear as relative positions without being brought within and subordinated to a larger space. In thought one might pass from one place to another, and remember the passage, but does not ascribe unique coordinates to any such place they have visited. The assignation (→1·2·3·←) thus recognizes that order is a way of packing objects as ideas within the boundaries of a given interest.

In a "transitive" relation we observe something similar to what we saw at 2, where a conditional relation was merely descriptive of the intervention of known into a field of unknowns, and 3, where a rule of extension was merely descriptive of the intervention of inference into a field of declarative formulae. A "transitive" relation likewise describes change. Two objects are placed in relation by way of a point of transition, such that "if a is related to b and b is related to c, then a is related to c." Points "a" and "c" are thus assigned location relative to "b". The point "→+" cannot occupy "·←" because that point is already occupied.

Reflexivity establishes closure and completeness on any given ordered field (set), such that each and every member of the field individually reflects and participates in the relation of order through which the field is unified. The process follows a rule that precludes an intermediate position for any event or object upon which understanding is achieved. An understanding applies to all members of that relation or they are, by definition, not understood.[13]

One may contend that this feature of human intellect sets the human apart from other species. The relentless compression of intuitively sensed insights into patterned theories marked by designs, rhythms and hierarchies generate the various abstractions upon which great human collaborative assemblies are based. It is not a tranquil process, as it constantly deconstructs and reconstructs human cognitive and social structures until the whole is understood.

Consider something as complex as a golf swing, which engages a field of possible motions, and which has a "+", i.e., flush contact with a ball. Within the swing there are a variety of symmetrical relations. There is a hinging and unhinging of the wrists, the vertical movement of the arms up and down, the rotational movement of the shoulders and hips, and the transfer of weight back and forward toward the ball. The swing is a relation of order, bringing a number of events, start to finish, into one idea. Those who understand the swing can often predict from the idiosyncrasies in the way an individual stands before the ball whether that individual will find it or not.

10. Yet as much as golfers toil over the reconstructions (refinements) in their golf swing, their labors lead to the formation of a single thing. There is a point where various patterns yield to sense of "wholeness" which settles over the project.

Look at a stuffed bear. We spend very little time and effort deconstructing our visual perception of the bear, and yet if we do this we will notice how the parts create subsystems—legs, arms, trunk, head, tail—and make a whole bear. Our construction of the whole is effortless, and without a need to deconstruct we accept it as it is, along with the numerous other objects of like "kind" which populate our "space." We just let it happen, and do so with a fairly workable sense of what fills the spaces we cannot see. Our experience of the world is structured with inferences that our perceptions acting alone do not report.

It takes a while, but at some point the mind becomes very good at carving spaces out of the seeming chaos of sensory data before it, and in constructing and locating physical objects within those spaces. For any given object, what one sees is a variant of a form within which one has become very familiar, a familiarity that enables the mind to wander outside pre-established limits.

At 1 we took those spaces as given, and attributed "true" to a space where interest integrated with a space so as to suggest agreement—converting · to (+). Having acquired some appreciation for the work that contributes to the establishment of a space, it is now apparent that a point "·" marked a place of transition between one order of magnitude and another—i.e., ending one and beginning another. One can begin to move logically from a simple particle to a compound extension bound within limits signified by "o".

The symbol o represents what one would refer to as an "acceptable" range of extension of variable systems of grouped places or "·"s. That space is determined, however, by a newly dominant interest reflecting a higher order of investigation. Interest has thus shifted from the ascertainment of + to ╬. Again, as in 1, ╬ occurs or is distinguished from o through the intervention of { }. One cannot exist without the other. Therefore { } functions as an activating principle intervening within a void of otherwise empty spaces.

At 1, we could not say where "·" came from, and the most that could be said about it was that it is a unit of comprehension. At this point we can offer a more substantive definition because we can say something about its origin, though the argument that supports that definition is circular. The lower order singularities are themselves undefined. The circularity is meaningful, however, because it suggests that development proceeds cyclically—and that is more than we understood previously.

To know the activating principle or "form" of a tree allows one to see and distinguish one tree after another, as special instances of a variable. Likewise, to know the form of an institution allows one to see and distinguish one after another, though perhaps with greater effort. They are each ordered (anti-symmetric) wholes, but the principle or "form" remains constant (symmetric) despite individualized variance. This helps to explain how it is one can assimilate into experience seemingly limitless diversity, without becoming lost in the process.

The depiction of the process at a capacity of 5 is the equivalent to that identified at a capacity of 1 except that each object in the field of possibilities is now recognized to be an item constructed out of lesser particles.

$$5\{o\} \qquad\qquad ooooo \doubleplus o$$

This depiction repeats the process identified as a point of beginning, except that it represents an ambiguity over magnitude. The expressions "·(+)" and "o{∓}" differ in that the latter includes two orders of magnitude, and thus conscious attention to that ambiguity.

First Order Logic addresses "object" in what it refers to as a "predicate calculus." This logic expresses ambiguity over scale—left reflecting completion and consolidation of a cycle, and right reflecting commencement and engagement of a new cycle.

First Order Logic (Level 5)	Developmental Logic (Capacity 5)
Predicate Calculus	Interest—new order { }
Universe/Term	o object (absence)
Universal Quantifier/Predicate	{ } interest (Next order primary)
Existential Quantifier/Negation	o to { ∓ } distinction
	Presence cancels absence

A similar expression of ambiguity is not present in the column pertaining to Development because it is not necessary. The elevation of scale is communicated in the sign, and the repetition in **5** of the information processing form of **1** is self-evident in the repetition of the form. In fact, the paired concepts in the left column serve as a prescription or recipe for the formation of an information processing schemata, as follows:

A "Universe" consists of a field of "terms," that if a term held the sign "·" could be expressed as "········ " or "ooooooooo".

A "Universal Quantifier" (as to say "for all") is a "predicate" assigned to an entire universe of terms and can be expressed as "( )" applying to "........ " or "{ }" applying to "ooooooooo".

An "Existential Quantifier" (as to say "for at least one") is a special instance in which a "predicate" applies to or distinguishes a term within that "Universe" and can be expressed as (+) or { ∓ }.

A rule of exclusion between a universe and a principle of quantification is descriptive of the alternative effect of presence upon absence—and does not need to be expressed.

As our thoughts range from one context to another, it is useful to identify the context out of which a given discourse occurs, and to call that context a "universe" of discourse. Within a first order logic such clarification is essential to the etiquette of communication. The situation differs in the logic of work, where mind is pressed against its own limits, and emphasis rests upon doing, and not just talking about it. The complexity of a task is thereby measured by the value that dominates one's plans. The measurement thereby derives more from what one seeks to do than whether they succeed or fail.

11. Our use of this notation has allowed us to view the assertion of "quantification" as a principle of collection or "membership" and that this assertion *negates* within a universe of discourse in the same way that interest selects or distinguishes as *true* an object within a field. A rule of exclusion, therefore, marks the difference between a world before and after the intervention of interest.

This approach allows us to avoid the "self-referential paradox," two versions of which are known as the "liar's paradox" and "Russell's paradox." The "liar's paradox" is evident in the statement "everything I say is a lie" which by including that statement means that something one says is true, which the statement itself precludes. "Russell's paradox" (named after Bertrand Russell for having written about it) is illustrated in "the set of all sets which do not have themselves as members," a principle of membership that cannot apply to itself since if that set is a member of itself, it is not included in the set, but since it is not included in the set, it is included.

This contradictory situation was disconcerting to Russell, as it should be. A contradiction such as this suggests that there is an error in the intended use of a term, and since there were only a few terms used—"set", "all", "themselves" and "member"—the error could be close to the root of the first order logic he had written.

Russell's solution to the paradox was a rule of exclusion, i.e., whatever defines or identifies all of a collection cannot itself be part of the collection. As our notation reveals, that rule is quite the same as the rule of assignment previously identified in first level logic: A or not A but not both, or by translation, that which forms or distinguishes a set (the function applied) cannot be a member of the set (its "product" or "argument"). That rule was an appropriate—though not entirely illuminating—solution to the problem. He was, in effect, recognizing the functional dissimilarity between two different types of things present in any given set. There is a "thing" that serves as its principle of membership (its intension) as well as the group of things gathered by it (its extension). Each exists for a different reason, just as a container exists for a different reason than its contents. According to Russell, the container exists at a different "level" than its content. The difference in level was a difference in "type" and he called this solution a "theory of types."

With this intuitive recognition in hand, Russell attempted to explain his first order logic with a meta-logic in which the things or objects conceived as "containers" could be organized into a hierarchy of "types." Self-referential paradox such as the liar's paradox and the Russell paradox could be thereafter avoided by clarifying the level at which a principle of inclusion is introduced. The *ad hoc* nature of this solution made a number of logicians uncomfortable, including one of his students, Ludwig Wittgenstein.

Wittgenstein argued that a hierarchy of types was unnecessary to establish effective separation between a principle of membership and membership. He pointed out that the separation between the defining principle (function) which gathers membership, and the members themselves (argument) was essential to the analysis of a set, or to any other object, and already represented in the symbols of logic—which separated "term" and "predicate." The symbols could be P(t)—the parentheses used to emphasize functional separation between the two, or in our case $\{\frac{1}{T}\}$—emphasizing that the separator and predicate (principle of distinction) are one and the same.

The analytic significance of "P" as a prototype or expectancy of membership is a different thing than "P" as member, or rather "P" in P(t) is different than "P" in (Pt), even though the same sign is used. Take "Dark" as a predicate through which "room" is examined: Dark (room), "Dark" in the first instance is not being used in the same way that "dark" in "that dark room" is being used. In the latter instance "P" is bonded to "t" as part of the object, but in the former instance is merely brought to "t" as the "sense" in which "t" is being examined. So, "yes," there is a hierarchy at work, but it is not an extrinsic hierarchy (functioning without the nature of a set or proposition), but intrinsically (within).

The solution in what we here refer to as a "developmental logic" or a "logic of work" the paradox is essentially the same, though perhaps more plain and obvious in the notational structure borrowed from G. Spencer Brown's <u>Laws of Form</u>. Here a state of interest (call it "sense," "function," "state of engagement" or "expectancy") is not a stand alone "object," but operates as a state of mind within which objects are selected as information. One is apt to encounter paradox if and when one attempts to objectify this state as or within a structure unless they first acknowledge that it in an important sense precedes structure. Any given interest may for a variety of reasons be objectified, but one must be careful to acknowledge that in doing so, they have changed what it first was. Upon first contact with a "field," one takes its objects as given, and then proceeds to select among them, and what we do when we select among ordinary objects is activated by a different interest than what we do with interests reformulated into objects. The hierarchy we have encountered is not "meta-logical," but rather "ontological."

In other words, the problem with Russell's solution to his paradox is the apparent failure to acknowledge that "member" exists only in relation to the defining principle of a set, i.e., that "member" includes its principle of membership. Therefore the treatment of a "set" as an object in a universe of discourse—a membership candidate—neutralizes the dynamic effect of its principle. The set as member is not what it was when it was a defining principle, or an expression of intention.

A reformulation of the paradox is evident in the expression "the set of all intensions." The set is empty because a member is, by definition, an extension, and is thus no longer an intension. But how could it be empty, if intensions are things? As one works through paradoxes of this kind, a realization emerges that in a very basic sense a state of interest or engagement is a pre-selective experience—essential to cognition, but not identical with its output.

This realization is more important than it may appear on a technical level. The appearance of a self-referential paradox is in fact the manifestation of a weakness in the proposition, "objects confer value on objects." An object cannot confer value on either itself or another object. Only value confers value upon something, and does so through investigation. This, in a manner of speaking, summarizes intellectual freedom.

Yet as we have seen, there does appear to be a hierarchy at work in the progress of thought from one level to the next. Thought progresses to larger more comprehensive landscapes, and it is therefore tempting to commit the error signaled in Russell's paradox, i.e., to validate forms of thought through the creation of a meta-logic, and within it to fashion a hierarchy of types such that there always appears to be a layer of greater elevation or depth to account for the others.

This sense of depth—where bereft of real interest—is a logically constructed illusion. The illusion can be revealed in the following exercise.

Take two points represented by "a" and "b" and combine them into a meta-point represented by A. Take two meta-points "A" and "B" and combine them into a meta-meta-point "aa". Then take meta-meta-points "aa" and "bb" and combine them into meta-meta-meta- point AA. How many levels are there?

This exercise suggests four levels of conceptual abstraction—a, A, aa and AA—because every combination of objects occurs through abstraction. This does not, however, resemble the kind of progression examined above in the logic of work. We are only seeing the repetitive application of a single process, which could go on indefinitely and thus produce infinite levels of abstraction.

The hierarchy which has unfolded under a developmental logic tracks methodological changes associated with the sequential coordination of mental activities, each representing a different state of engagement or interest. There are only a few of these brought to a field in the course of a developmental cycle, and they are brought with difficulty. The landscape changes as a result of motivated interventions, and becomes broader and more comprehensive—because the integration of skills allows this. Increased dimension is more an effect of this integration than the substance of it.

Importantly, the changed landscape involves altered contours, as increase in the shape of experience is bound to its magnitude. These shapes are supported by shifts of motivation, and the flow of time in the definition, passage and closure of events may now be reformulated in terms of the intervention of value.

As simple as this distinction is, it is remarkable how theorists of organization tend to conflate logical and abstractive hierarchy. The "science" of organization that emerges from this deficiency leaves it without an effective calculus of description. The resulting view is a mind bound within structures instead of one that transcends those structures through motivational affirmations. An accounting for such movement is therefore offered as a preferred description of the free intellect in which healthy organizations form.

12. Because an ending to development within a context of one magnitude has been identified as the beginning of development in another, it is possible under this model to conceive of a seamless process of growth starting at infancy, and proceeding through childhood, adolescence and adulthood. It is also possible to deploy a more discriminating set of measures in the determination of the progress one makes. There may

be an appearance of growth when it does not exist, as is the case where one persists in growing an institution or some other fortune for the simple pleasure of making it larger. For that matter, one may have grown even though there is little outward appearance of the same, where his expectancy has shifted to something higher.

Once engaged in real learning, a certain space opens to the intellect, a developmental space whose measure at a given moment is marked by something which has been called "capacity." As in space, size is known both by the perception of barriers and the passage beyond those barriers. These have been characterized as rules of exclusion and the thought that overcomes those barriers. While examining them, it is possible to fasten attention to the lexicon of formal logic, and to redefine it with descriptive pictures. In the process, it becomes steadily more apparent that there is little to distinguish the formative behavior of an infant from that of the director of a complex institution, except for the character of the expectancy that motivates and supports the intellectual endeavor and the mental foundation that allows one to take matters already learned for granted.

The only state of consciousness for which there is no rule of exclusion is infinite capacity, because it, by definition, has no limit. If the infinite is represented in a symbol * then Reason, the boundless region into which intellect grows, i.e., manifests freedom, could be represented as well.

$$*(*+)$$

This is provocative, because * is presumed here to have Interest by virtue of ( ) embracing *+. One then contemplates whether Reason is indeed value free, and if not, what Interest extends from infinite capacity, and what the existence *+ is which it embraces as Truth.

If this is what Reason is, then it may, from an intellectual vantage point, be quite difficult to imagine. This is not because we are incapable of knowing some or part of what *(*+) means. What we are fully capable of is not known. It is likely that if one were for a moment to be vested, as a kind of visionary, with a sense of this infinite "idea" they might report to others that it wasn't much like the "reason" imagined by the scientists and philosophers of the age.

There is a good deal of prose devoted to the science fictional notion that machines may someday acquire an intellect that interacts with human beings in a way that reflects an achievement of "self-awareness."

This chapter's examination of the human intellect, however, belies this comparison. Since the functioning of intellect presupposes interest, a motivating expectancy, the machine, though simulating interest based activity, will lack interest unless it can want something. A machine that wants something is, by definition, not a machine.

Intention is the basis of metaphor, and assuming that there are mentally fixed programs which lead toward the formation of objects, their expanded deployment into diverse and unfamiliar contexts is a metaphoric extension. In order to say or perceive that a given operation has analogous use within a new field, one must be able to say or perceive "x is to y as a is to b." The portion of that expression "is to" is an expression of intention. There is no mechanical device which, absent the experience of value, can effect such a change.

But then again, one may choose to behave like a machine. Much depends on the way one sees oneself. The ideologue and the sophist both adopt a somewhat mechanistic view of themselves, one which emphasizes its incapacity to error, and the other which emphasizes the irrelevance of error.

3.

Moral Freedom

The shortest and surest way to live with honor in the world is to be in reality what we would appear to be; all human virtues increase and strengthen themselves by the practice and experience of them.
Socrates

1. Despite the achievement of some agreement among those few who toil over the issues of logic, the audience that benefits from their work is small. By contrast, the entire human race is prone to chatter over matters of right and wrong. Values, it seems, are quite important to us in daily conversation, and live among us in the stories we tell.

The examination of intellect is an examination of what one does in the pursuit of interest. Though mentioning "interest" frequently, the discussion of intellectual freedom made no effort to discuss whether a given interest is good or bad. Moral freedom, on the other hand, engages and evaluates the content of human interest, and how we come to distinguish moral from immoral objectives.

One may have thought that we have unlocked the science of organization by identifying its descriptive calculus, in effect, by recognizing that cohesive organization (object) is enabled and completed through the intervention of interest. Having identified a rule of motion by which objects are selected, we are now at a loss in understanding how interests are selected—or whether to speak of interest as a process of selection is in any sense intelligible. One might blunder into this topic by reducing selection to instinctive motivations, but this is apt to have a disastrous affect on social relationships—whether conceived as a hunting party or an internationally scaled institution. Interest is the result of something more complicated, and cannot be addressed coherently without considering a process that we think of as a "moral" process.

As one might already have gathered, the terms "moral" and "ethical" are not identical. There is an ethical component to logic. Logic cannot exist without the intervention of interest, and without a prescription of better or worse that derives from interest. However, the logic we regard as useful in work systems, and which accounts for institutional hierarchies, focuses on relatively stable objects—stable in the sense that there is some social consensus about what they are. Morality, on the other hand, focuses on a less stable experience—i.e., an individual's relationship to society. So, although one is likely to see resemblance between the ethical content of logic and of morality, they are not likely to express themselves identically because they engage different subject material. Morality is a subclass of ethics, an ethical affirmation, adding to interests we think of as "desires" or "appetites."

Take a primitive "interest" as in an interest in survival. This interest might be explained through its function, i.e., to preserve genetic formations, but this explanation is, by all counts, superficial. It does not reveal the content or experience of such an interest, because it does not do more than identify a functional relationship between an experience of value and its effect—wanting to survive promotes the continuation of genetic patterns. Similarly, *survival* fails to reveal the content of one's interest in food. The experience of this desire is simply a matter of feeling hungry.

Since "survival" is not a desire like others, and is not typically part of the experience of desire, it reposes trust in the selective process of genetics to program a given species to want what is good for it. Yet wanting to survive, and valuing the life that survives, is arguably a higher order interest than simple appetite, and calls attention to a number of difficult questions addressing the "what" and the "who" which survives.

At least two factors influencing the establishment of interest appear in intellectual freedom. First, consciousness is capable of establishing interest of sufficient weight and magnitude to engage its complete effort and commitment. Second, it is apparent that a given interest is something that someone can grow out of, in order to embrace something more meaningful. The fact that a given expectancy can be superseded by another is evidence that interest tends to include more within it as it advances through or within a hierarchy, but whether it is more of a good thing, or a bad thing, or some combination of the two, cannot be determined without adding a point of reference that allows us to make such a distinction.

To some degree, it helps to be intelligent in the choice of what one's interest is. Also to some degree, these choices assist one in being intelligent. That there is a connection between them invites the argument that smarter people are morally superior. The flaw in this argument is similar to the error of supposing that smarter people manifest greater intellectual freedom. Unless one chooses to place intellect in the honest service of interest, the intellect is not free, and cannot progress freely even though one is very smart.

Similarly, one must examine interest to determine if it follows a force of mind—as intellect follows interest—that dignifies and liberates interest. In this chapter on moral freedom, the force of mind which uplifts interest is one's identity or "self."

Some might argue that the opposite is true, and that morality is a state of self-denial or negation. Altruism is generally regarded as an example of moral sentiment, as action designed to benefit others embraced by social systems, and is regarded decent and virtuous. But then again, perhaps the "sense" of self which promotes altruism is uniquely elevated and honorable. Indeed, whenever we find ourselves evaluating good deeds, we do so in reference to the goodness of the person—the character or moral identity of the person who performs them.

As commonly as the word "self" is spoken, there is much uncertainty over its meaning. If there is such a thing, then what is the entity that speaks of it? If certain interests are "selfish," what is it that chooses to act or not act on those interests? If one's "self" chooses not to act selfishly, then was it accurate to refer to those desires as selfish? The selfish interest may not represent a self at all, but might have been posed as such by true "self" in order to help clarify and weigh the alternatives. There is a self present in the statement "I decided" which may be a truer self than that which was supposed.

Just as in the study of intellectual freedom, we refrained from measuring the goodness or badness of interest, likewise in the study of moral freedom, we do not measure the goodness or badness of the self or identity that promotes an interest. We are instead preoccupied with whether and how it stands upright in the face of social demands. The moral self is real and actual, or to put it another way, it "exists."[14]

Accordingly, we approach the evaluation of persons with strong and resilient egos circumspectly. Moral reproval is reserved for those who have not undertaken the challenge to form a viable social identity, or who have permitted their identity to be displaced by others. This explains why, under certain circumstances, one may have greater respect for a bank robber than for one living his or her life in timid compliance with a crabbed view of the perfect life.

While addressing the subject matter of morality, it is useful not to overreach its limitations. In fact, an error that seems to repeat in modern civilization is the use of moral condemnation to address behavior which is quite moral, though on other levels reflects political and/or spiritual depravity. As one comes to understand the basis of moral sentiment, they may judge behavior as morally upright and integrated and yet—for political or spiritual reasons—ethically condemn it. Moral freedom reveals, among other things, the ethical limitations of moral sentiment.

2. One's best interest is evaluated in the context of a society of interests, some of which are in competition, and some of which collaborate. As the Socratics recognized long ago, competition and collaboration occur within a person, and between persons. When one considers morality, they consider social norms, and the arguments in support of compliance with them.

The subject of moral freedom addresses that state of mind which supports compliance with those norms. Some comply begrudgingly, and would not do so if there were not a socially imposed penalty associated with non-compliance. The more common experience of moral rules, however, involves a more intimate relationship between those rules and one's selfhood. The mental state of a morally free individual is one that embraces and promotes the establishment of norms which serve to define and express who he or she is. Morality is not a matter of avoiding punishment, but of desiring and acting in conformity with a viable social identity.

There is some resemblance here between this view and that of an evolutionary biologist, who would emphasize how genetic information is advanced by strong individuals nurtured and protected by strong groups. One can certainly imagine how, over many generations, our brains evolved so that we would be motivated to combine personal and collective objectives.

While there is a plausible connection between a conscious experience we refer to as an "identity" and the continuity of genetic information—via survival, mating, family organization, etc.—there is yet a difference between existing as an ego and existing as a biological organism. They are not the same thing, and may at times be incompatible. There are times when we may prefer compliance with a moral rule to survival, quite possibly because compliance with moral rules is required for the type of existence we call "ego" or the "self."

The promotion of genetic information is, in fact, an erroneous standard by which to measure moral value. Imagine two species, one aggressively cannibalistic and the other not. There may be circumstances in which the cannibalistic species survives and the other fails, but that only records a consequence of such behavior. One would think that, in general, species which depend on social cooperation would not fare well with cannibalism. But here, the moral calculus is quite complicated, as there are social species that control overcrowding through cannibalism. In truth, we do not know whether moral behaviors will ultimately kill us or not. And what benefit are moral rules if they do not inform of the wrongfulness of conduct before it happens?

It is sometimes amusing to see comparisons in popular literature between human beings, as a species, and cockroaches. The humor derives from the suggestion that since cockroaches have a better chance of surviving off into the distant future than human beings, they may be regarded in that sense as the superior species. But the point of the comparison is that species longevity is an irrelevant consideration in assessing the ascendant moral value of a living being. Moral value is barely applicable to the understanding of cockroaches, if at all, but highly applicable to the understanding of human beings.

The benefit of having the perspective of evolutionary genetics available to the evaluation of moral rules is that it allows one to be pragmatic in the application of such rules to a decision about how to act. This might be applied by evaluating options with a system of weights and measures, which include the practical value (survivability) of socially stabilizing rules together with rules that tend to preserve personal choices.

Such pragmatism, however, if governed by the goal of species survival, acts as an extremely blunt instrument in the evaluation of moral norms. Species survival tends to support rules that favor collective interests over individual interests, particularly where decisions are made under informational or time limits. Moral rules are, by design, subject to exigency that precludes remote calculations. We would, by adopting standards testable only over millennia, have placed ourselves by default under the domination of very crude collectivist norms.

We have noted persuasive attempts by Plato and Kant to explain moral behavior as a product of reason, as the only faculty of mind capable of displacing personal interest in favor of a larger and more coherent order. We left that discussion with unresolved concern over whether such disinterest effectively explains the sense of commitment associated with moral behavior. That concern is hardly lessened since our examination of intellect reveals that intellect cannot function disinterestedly.

If one admits that the interest which engages objects is, in effect, a choice of one's identity—a self serving choice, the issue merely shifts from the content of interest to the content of selfhood or ego. We must allow the possibility, therefore, that the formation or presence that we refer to as "self" includes "other." It is unnecessary to imagine a "reason" or "intellect" that stands apart from desire and tells us what to do and what not to do. What is more important is understanding how "ego" and "other" become unified.

Moral interest is ego seeking a way to exist within a surrounding social group. Fitting in is not a trivial preoccupation, as one needs the group to live. It includes a need to make claims within—and at times against—the group to satisfy one's needs. The substance of moral interest is an interest that marks a state of compromise between collective and personal welfare. There is no rational detachment about it at all, but to the contrary, the formation of moral identity is very emotional and personal.

As much as we are programmed to fashion this identity, we are similarly motivated to encourage others in our group to do the same—and for the same reasons. The group assists this process. Norms which accomplish this are transferred between generations through commands, instructions and stories too pervasive and numerous to resist. These norms are not just a matter of recollection, but are appropriated into and extend from one's identity, as an entitlement or need.

Thus assuming a socially integrated ego brought on by thousands of generations of human existence, and by social cues and pressures too numerous to count, we are nonetheless challenged to deconstruct this ego so that we might identify principles common to all persons—of coexistence between one and many. The task, then, is to understand the continuity that exists, if any, between a moral identity and moral behavior. That, in effect becomes the standard by which moral rules are judged, i.e., by their tendency to support and sustain one's social existence such that through them one can become whole and complete social being.

Whether this being is more or less likely to pass on genetic information—as a biological organism—is of little practical or theoretical concern. Having supposed a socially based conscious experience called "selfhood" or "identity," any curiosity directed toward genetic mutation takes a back seat to the mental constituents of that experience. Even assuming physiological structures make the discernment of "soul" possible, moral freedom examines the experience from the inside to give an account of what it is.

Within the rational universe of evolutionary genetics, in fact, once selfhood proves to be an effective promoter of genetic material, natural selection would tend to serve or enhance that experience. In that sense, even within the genetic universe, the "self" acquires a life of its own—an organizing principle affecting our physical instrumentalities.

3. Allow that the human is a social creature who depends on and collaborates with others for survival. Normally it is in one's most "selfish" interest to be accepted and trusted by others who participate in the

preservation of one's welfare. While this provides a reason to behave morally, it may not be the reason for moral behavior. As stated above, mind may have evolved to a point where, in combination with extensive societal cues, the embrace of a moral rule is pre-rational, i.e., automatic and reflexive and not the result of contemplation.

To put it another way, moral choice is a matter of identification rather than strategic deliberation. Moral interest reflects and is actuated by selfhood, and is thus a manifestation of the successful formation of ego.

While many influences bring about the formation of an identity, identity is the platform from which one chooses among values, and what is right to do. It is in that sense autonomous, not because it is free from influences, but because the establishment of a moral perspective determines the manner in which those influences will combine into a choice. Society is not an overlord which determines selfhood, but the challenge out of which a self emerges and fashions moral interests out of the chaos of desires.

Let us then consider what this emergent presence is. The "self" is not a thing in the sense that a tree is a thing, nor is it a thing in the sense that a desire or interest is a thing. It is a different kind of mental phenomenon, even though it is present in and the basis of moral judgments. It exists in a way which precedes structured articulation, and is experienced most acutely when choices are difficult. It is like the inner sense that provokes the artist or composer to choose lines and colors, notes and tempos.

One's involvement in a society is a creative endeavor. Until the challenge is undertaken, selfhood is more like a shadowy potentiality, a nonexistence seeking existence. These choices must, by the nature of the undertaking, accept that competition over food, shelter and sex is real. One therefore looks for ways in which they can satisfy those needs while limiting the social conflict resulting from their indiscriminate pursuit. They become "goods" that help to define who and what one is.

Consider the two institutions of property and marriage. It is probably correct to say that in order to eat and to be safe from and comfortable among natural elements and predators one must labor. However, the incentive to toil is weakened by the prospect that others will consume the product without offering anything in its place. It is likewise acknowledged that sexual gratification has, at least until recent history, involved the birth of children, the personal and social interest in caring for them, and the formation of families. However families are undermined by sexual infidelity.

Typically when one is asked who he or she is, they will start with a statement somewhat like, "I am a worker, an owner, a spouse and a parent." All of these ideas are imbued with moral value as a claim against others supported and defined by social consensus.

In modern civilization there is a "golden rule" stating that one should "do to others what you would have them do to you." This rule has no meaning without the affirmation of interest to which is ascribed "right." If the rule meant that one should refrain from doing anything they would not want done to them by others, then it would enjoin conduct that reflects competition over things in limited supply, or that otherwise represent interference between or among diverse interests. The golden rule is unbroken, however, if one commits an act which, though interfering with the desires of another would not begrudge similar conduct by the other.

It is the "would not begrudge" part of this formula that calls thought back to the idea of a right to which the self-expression of desire is affixed, i.e., there are behaviors that are a perceived "good," even though they occasionally interfere with one's own desires. One may own property next to their neighbor, who builds a house blocking part of a pleasant view, or intrudes on areas of privacy, and yet something about "right" associated with "property" causes them to yield, voicing no objections. Under right of property one can claim what they desire and know that it will not be the source of much, if any, social discord.

One might imagine a utopia, a society which somehow organizes human desire in complete accord with a line of supply from an omnipotent source. Suppose a super-technologist were able to fabricate a convincing, though fake, mechanical satisfaction for all sexual fantasies. Would the institution of marriage continue as it was?

The point is that the inner presence we refer to as "self" connects to a social environment through behavioral standards. Strong needs to restrict behavior include a self that demands more coherency. In the case of a nomadic culture that risks disintegration through contact with many societies, one would expect a unitary and cohesive form of identity, represented by a singular and exclusive deity such as that formed by Judaic tribes which roamed the Middle East thousands of years ago. Emerging from a highly assertive and cohesive sense of identity was a set of commandments which established norms of worship and behavior which continue in and are important to post-industrial societies.

But there have been significant modifications in the application of behavioral standards. In a technologically advanced society, diverse market economies and abundant food supplies have altered the importance of behavioral standards upon which pre-industrial civilization depended for effective self-expression. The removal of social and physical obstacles to personal satisfaction has compromised institutions to which many older cultures have become completely identified. When these cultures are challenged by these new realities, it is more than a matter of adjustment to new habits.

If moral identities and the values extending from them were nothing more than strategic choices, it would be difficult to explain the volatility of attempted communication between social and moral types. The estrangement between such groups seems more like a failure of recognition—an inability to comprehend the behavioral norms put forth by a different group.

Normative structures promote the peaceful and trusting relationships necessary for effective social collaboration, while allowing the satisfaction of personal need. One does not choose them as one would in weighing the benefits and detriments of competing alternatives. One's hands have a strategic value, but it can hardly be said that one chooses their hands. They are an extension of what he or she is, as an organ of sense that connects them to the physical environment.

This society in many ways delivers these values as standards by which one fleshes out a self-image that may or may not be positive. The loss of existence is arguably the phenomenon to which Kant and Plato were directing their attention when they proposed that a rational faculty is the cause of moral behavior. The violation of the norms of social self-expression, such as the transgression of property or marriage boundaries, cannot occur without a loss of one's sense of existence, because the transgression itself weakens one's sense of connection to others.

From this it is plausible, if not indispensable, to think of "self love" less as a value that promotes one's own desires in opposition to other, but as a state that seeks to include other in the formulation of self-interest, for without such inclusion the self on which our selections rests tends to disintegrate. In this sense, self love is evident in its moral affirmations. The recognition of this is the substance of moral freedom, and the avoidance of it is very similar to what a state of moral enslavement is like, where personal appetite functions as master in search of servants.

4. We gather with others because we are social creatures. Being, along with the health of such being is bound—in mind and body—to a

group, and we find ourselves within that group. And yet we are uncomfortable around those who ask that we surrender our existence to a greater cause—their cause, of course. Just as we once learned to walk, the task of balancing self-affirmation and self-denial challenges us, and requires practice and attention.

Though we condemn as immoral individuals who refuse to appropriate the behavioral norms of a given society—the criminal or sociopathic deviant—we do not measure out our expressions of moral approval in proportion to the intensity of one's identification with those norms. To the contrary, over-identification with those norms is a perversion of a different kind which we regard as morally loathsome.

What is our revulsion toward the sociopath? It must be that there is soullessness to it, a light which fails—no empathy, no recognition in them or by them of shared values. There is intelligence, but it is bestial intelligence. It is something we sense in them, something we look for and find lacking. Their moral practices are naught but strategic positions, and reveal little of who they are.

But we sense a similar lack in the one who surrenders their personal identity to the group—the moral zealot. They just aren't there, though they do all that is expected of them. This life is in a state of constant compliance, completely preoccupied with states of moral approval and reproval—and distributing it in like measure to others. And we accept such persons because there is so little to reject, and we avoid them nonetheless because there is little there to like.

Both perspectives are immoral because the essential task of morality—the formation and maintenance of a socially constructed identity—has failed. Moral freedom, by contrast, recognizes the collaborative significance of selfhood and interest, where one takes the discovery of their identity seriously.

An aversion to the sociopath ordinarily wanes as we gain control over it. Despite the predatory intentions of a sociopath, his or her predations are subject to containment by organized coercion. Such personalities, once contained by their own fears, can make amusing companions, provided that they are handled cautiously.

The moral zealot is, in many ways, a much more formidable social adversary. The surface persona tends to attract moral admiration, but fails to attract friendship. It speaks of moral absolutes which are difficult to adapt to human needs, and insists upon recruiting others to its viewpoint. As the number of followers grows it is apt to gain influence over the institutions which make and enforce laws, and treat peaceful nonbelievers very disrespectfully.

Extremes represented in zealous over-identification of this kind often manifest in compulsive hyper-achievement. The results can be impressive, especially in post-industrial societies such as ours, which reward the winners in contests of skill—and of other competitions—with admiration and wealth. These can be quite tempting, and one might fall into a state of sacrifice in one heroic spectacle after another.

Their behavior isn't particularly immoral, but it fails to resonate in moral self-expression. Such resonance and a developing receptivity to increasingly potent modes of moral satisfaction are the direction which brings one out of the daze of self-exultation and importance which often attaches to celebrity. Some achievers are better at overcoming the temptations that this presents than others. Competition has a way of sorting out authentic performers—those who love what they are doing—and those who are engaged for the social image it casts.

It seems therefore that useful in evaluation of moral behavior is the acknowledgment of a universe of behaviors that tend to attract social censure or approbation, and in ways that are ambivalent or conflicting. Giving oneself away to social approval may lead to being buffeted by change among groups and circumstances, because there is no self that identifies with behavior which conforms to a social virtue. This might account for our aversion to the celebrity of a politician who vacillates on principle in order to be liked. It also accounts for the nuance of our moral examination of wealthy individuals who—in addition to living life in comfort—seek social validation of the choices which led toward the accumulation of wealth. Have they become lost along the way?

Perhaps they have, but the word "lost" might be too strong. Such individuals seem to know exactly where they are and what they are doing, and will say so without much encouragement. All it takes is a little patience, and they will tell a story about their life—a story of "success." But for most, there is an experience of divergence between socially formulated ideals and a more personal experience which seems to appear as the result of the assertion of moral autonomy.

One can imagine intellectually brilliant people as participants in moral failure because the choice that produces moral freedom represents more of an orientation to one's identity than to any given object. One may command great armies and industries, while their identity eludes them. However complex an object may be, it is a relatively stationary entity compared to one's self. Therefore despite what theorists may imagine as a concord, of sorts, between intellectual and moral accomplishment, there is no necessary adhesion between them.

Whether a moral person is a bigger person is not an easy question to resolve scientifically or philosophically. Much depends on the establishment of playing fields in which moral rules are suspended or altered in order to promote the vigor and creativity of its participants. In what many theorists now refer to as "games" there are substantial moral ambiguities, such as in a competitive industrial marketplace, where fairly craven behavior is heralded for its innovation.

A society of that kind—a manic and paranoid world of individuals pushing each other aside in a scramble after produced goods—hardly resonates as a true or real state of being. Nor is the situation appreciably better where groups of individuals surrender their moral autonomy to local or tribal normative systems. Tribe against tribe is only slightly more dignified than person against person.

This state of moral error—which excludes genuine personal engagement with the collective—leads toward loneliness and isolation, like a person sitting beside a pool while the others play water games. He or she misses some of what it is to be a human being. Social engagement that includes emotional immersion in the collective is a different kind of choice than we saw in the developmental cycle of intellect. As such, it is a different kind of freedom.

The difference does not, however, draw one away from human institutions, and the way they are put together. Much to the contrary, it brings one face-to-face with one of the paradoxes essential to the design of humane societies. One might attempt—in a manipulative way—to encourage individual surrender to collectivist norms, but at some point this is apt to undermine the collective as individuals attempt to break free. What sort of resistance are we likely to encounter? What sort of compliance pressures are necessary to get people in such environments to do what we want them to do? As these questions are addressed, it is possible that we will arrive at insights focused on a relationship between an intellectual creativity widely regarded as productive and beneficial and moral autonomy. While it is not necessary that moral and intellectual freedom adhere, it is conceivable, nonetheless, that they may—though not through force.

Whether one may, through manipulation, influence others to adopt a proposed social norm as moral is questionable. As the ensueing discussion will reveal, such manipulation (usually not a good thing) has less to do with one's control over the facts than with their control over the fictions in which facts are embedded.

5. Even though cognitive ability is a poor explanation for human compliance with moral norms, there are significant cognitive elements to such compliance. In the behavioral science of mental disorders, one cognitive trait that separates sociopathy from virtue is an inability to achieve empathy. Scientific theory might then suggest that a lack of care for others follows from an inability to mentally reconstruct the experience of others.

To say that caring (interest) follows after reconstruction (object) is, as previously demonstrated, untenable. Given that interest seems to initiate what we think of as learning, it would seem that a cognitive change does less to explain than to mark developmental transitions.

The use of descriptive matrices that emphasize a developmental aspect for the formation of socially oriented norms strongly implies interplay between one's cognitive awareness of others and the affirmation of his or her value as a person. Take the following developmental paths—mentioned in Chapter 1—comparing social behavior and social interest. We will refer to this as a "social science" approach to moral value.

Kohlberg (Social behavior)	Maslow (Social interest)
1. Sanction (Personal orientation)	1. Physiologic (Personal orientation)
2. Reciprocity	2. Safety
3. Altruism	3. Intimacy
4. Law	4. Status
5. Justice (Social orientation)	5. Self-actualization (Social orientation)

That behavioral science seems to have created a divide between what are characterized as moral rules and personal need reinforces the contention that there is a primitive conflict between what one wants and what society permits. However one can, by contemplating these pathways side by side, appreciate that there is a unifying element of some kind which functions among or between these developmental pathways. It is, for example, difficult to avoid noticing that both columns reveal the incremental growth of a socially inclusive frame of mind—i.e., from personal isolation to the internalization of socially unifying principle.

Moreover, there appears to be one-to-one resemblance between increments one through five in each column. The morality of Sanction, i.e., choosing to act only with reference to reward or punishment associated with that act reflects the disposition of one concerned with "need" at a primary level, that is, the attraction to pleasure and the avoidance of pain. The morality of Reciprocity, i.e., giving favors with the expectation of a return, involves the use of contract to control a socially unpredictable environment.

The morality of Altruism, i.e., ascribing value to another, reflects a need for personal validation or intimacy. The morality of Order, i.e., ascribing value to a system of rules which give priority to certain interests over others reflects a need for status, i.e., the simple understanding of where one stands in the social pecking order. It was Plato who advanced the idea that Justice and Self-Actualization are synonyms.

While working to comprehend the unifying feature of these developmental progressions, it is useful to remember that the words have been chosen from the vocabularies of different theorists pursuing divergent interests. That the two columns may, in fact, derive from a common source attracts theoretical inquiry toward the content of that source. This attraction increases as we add to the collection of seemingly related systems of value.

Below it is apparent that social science straddles two different worlds—uniting on the one hand with cognitive science and theories of logic and on the other with moral value—such that it is often difficult to determine whether social science is descriptive or prescriptive. While social science is descriptive of normative reasoning and motivation, the developing "improvement" visible in normative experience strongly implies an ethical principle at work in this development.

1. Cognitive Science		2. Social Science		3. Moral Value	
Intellectual Behavior	Intellectual Interest	Social Behavior	Social Interest	Moral Behavior	Moral Interest
Declarative	Primary	Sanction	Physiologic	No killing	Existence
Cumulative	Retention	Reciprocity	Safety	No Stealing	Property
Serial	Extension	Altruism	Intimacy	No Adultery	Trust
Parallel	Order	Law	Status	No Perjury	Honesty
Declarative	New Primary	Justice	Actualization	No Coveting	Autonomy

The columns placed under "moral value" verify not only a progressive movement from lower to higher states, but that each phase of growth carries its own resonant moral value. While we can describe what moral people do (science), it is quite another thing to examine behavior and say that it is a right or wrong (ethics).

Perhaps the most important point revealed in the extended presentation of developmental sequences is that descriptive terms of normative experiences appear to be extending from a central presence that includes intellect but is not dominated by it. This entity bestows value, because it exists to bestow value or care. Certain moral principles extend from this presence. Reason does not make them so, but is part of an integrative process which reveals their value to us.

Moral freedom thus argues for the gradual emergence of socially reinforced selfhood, and the establishment of normative interests that serve that existence. These achieve sanctity because they establish a sanctuary in which selfhood is experienced. Without them, the domain in which self can be said to "exist" is destroyed and one becomes a mutating field of force under constant vigil from some unknown "I."

Whatever validity may be attributed to the claim that a certain kind of existence is better than another—which invokes political and spiritual issues—such advocacy is meaningless until existence itself is established. A comparison of identities cannot occur unless identity is first established. A being, whose consciousness only serves as a receptacle for pain or pleasure, success or failure, satisfaction or dissatisfaction, and the cognitive labor activated in pursuit of such things, cannot answer the question, "am I happy?" One cannot be an "I" who is happy until they are an "I" who "is." All that exists is a subject attached to an inventory of desires.

6. As the preceding discussion suggests, it is difficult to be precise in the description of moral value, and a self that engages such value. The problem was less apparent when we examined the intellect and the structuring of wholes. That discussion examined a relationship between a highly structured "object" and a comparatively unstructured interest.

Here we are concerned with an entity or "presence" which is even less structured than interest and which we suppose to operate as a growth medium to moral interest—the socially engaged ego. How do we show what this is?

Our difficulty in the use of language to express moral interest strongly suggests that more is involved in sharing such an interest with others than is involved in the use of words to denote common objects in our experience. We can with a word such as "altruism" identify a motivational state that others may or may not be familiar with, but we do not express the value of that experience. The value of the experience may not be important if one is having a purely academic discussion about moral virtue. But the communicative task changes considerably in a social setting where there is a need to promote or validate such an experience.

Words or other signs alone may or may not be effective in stimulating one to experience or recollect states of value. Values as simple as pain and pleasure are not like the objects of a shared "exterior" world that one can identify with a gesture. If someone says "it hurts" another is apt to ask "how does it hurt?" To clarify "it hurts a lot" does not advance communication appreciably, unless one identifies an experience that may be common to another, such as to say "it hurts in a way similar to the experience you may have had when boiling water was accidentally poured on your skin."

The description of physical pain is more challenging to express than the moral value represented in the statement "I feel guilty" in part because such feelings require an articulation of a complex of nuanced circumstances involved in the formation and violation of moral value.

Since such values are not effectively communicated through a simple word or precept, we tend to embed those values within a more elaborate arrangement of the circumstances, likely including a variety of familiar personal motives and social constraints. These circumstances often appear as personal narrative—stories—which organize information for the purpose of evoking that experience in others.

In complex societies, these narratives may fall into familiar patterns that are, not surprisingly, quite popular. The various writers, animators and actors of these narratives combine into entertainment industries. More traditionally, important narratives are told and retold at religious gatherings. When we observe these narratives in primitive societies, we call them "myths," a somewhat condescending term given the energy and enthusiasm with which we consume a daily diet of emotionally engaging fictional descriptions of our own lives.

These stories may or may not have a factual basis, but in every instance involve a distortion of truth in order to highlight a moral value. In the moral life one is apt to see their own experience as a story unfolding.

A myth is a medium through which moral value is articulated in the same way that logic is a medium in which intellectual value is articulated. Having already noticed a developmental component to moral value, one may anticipate resemblance of form between logic, pure form, and the structure or "plot" of stories. What is noticeably missing from logic is the presence of details aimed toward the stimulation of moral perceptions or values. What is noticeably missing from "myth," on the other hand, is the preoccupation with factual truth. Exaggeration is common in mythical narrative because its purpose is to evoke a motivational state.

In logic, one takes value as a given. In morality, interest is the dominating subject-matter, but is less accessible for examination than a structured object of perception. Even if one tries to examine their interest while conducting an inquiry, that examination will be elusive. The interest that drives investigation cannot look with detachment upon itself. An attempt will likely not succeed unless one is able to move into an entirely different point of reference—thereby disrupting the investigation in its original form.

Generally speaking, moral reform and progress is hard work, occurring in mythic transformation through the acquisition of virtue. Through virtue one does more than adhere to a socially viable norm, but identifies with that norm as an expression of who one is. In "virtue" there is no differentiation between personal and social affirmation. One neither abandons selfhood to a moral code, nor views such a code as a constraint upon self-assertion.

The difficulty of achieving virtue is equal to the difficulty identifying with ego enriching values. Reason does not, for example, confer value on life. Life has a value that resonates with consciousness when given the opportunity. It is not about figuring out virtue, but of enabling the experience of it as we tell our story to ourselves and our friends.

Since moral value is arranged in a developmental series, one would expect to see corresponding narrative. Here we take the classification of mythic narratives found in Northrop Frye's *Anatomy of Criticism*. There is an interesting correspondence between those narrative forms and the information processing forms which describe the successive tiers of intellectual development.

Intellectual Processing	Intellectual Interest	Mythic Narratives	Moral Interest
Declarative	Primary Interest	Comedy	Existence
Cumulative	Retention	Romance	Property
Serial	Extension	Tragedy	Trust
Parallel	Order	Irony	Honesty
Declarative	New Primary	Comedy	Autonomy

As we proceed with the examination of mythic narratives—and their emergent versions of virtue—one cannot ignore a resemblance between the story patterns of these narratives (plot forms) and the information processing patterns found in a developing intellect. As mentioned earlier, while there appears to be similarity in intellectual and moral development, the resemblance does not include a co-dependency between them.

The search one makes for oneself, a true and autonomous social being is—despite important similarities—a different kind of search. One's identity does not sit well among common objects. This should become clearer as we look more closely at the stories we like to tell.

7. So begin with Comedy.

Life is not a value if it, as a word, is meant to call attention to the fact that someone is alive. But when we make a point of saying that someone is "alive" or that they are "full of life" we usually mean it as a compliment. By way of this virtue a given individual takes up space, demands to be noticed, and exhibits comfort with the maintenance of a social presence.

Such existence marks a point in which a person assumes membership in a social collective, and is subject to moral judgment. For a period of time in one's life, society holds moral judgment in abeyance, as a child prepares to become an adult. A "child" cannot be guilty of a crime. A "child" is judged in terms of whether they please adults, and many who should be adults live out their lives seeking grownups to please. This amounts to a moral failure, of sorts. But one might also seek a way to act without having to justify oneself to others, and then a Comedy ensues.

In Comedy, the main character is confronted with one predicament after another having no particular order or necessary sequence. Though not necessarily humorous, the pattern itself attracts humor for a number of reasons. The hero or heroine is a novice who, in an innocent way, often appears foolish. Their foolishness debunks an authoritative figure which lives to restrict the actions of a main character and his or her compatriots. The frustration of authority combined with the giddy evasive maneuvers of the protagonists resonate with the joy of existence and laughter ensues.

This shares the pattern of the first stage of logical development. At this stage one forages among the possibilities in no particular order, a situation necessitated by the fact that all of one's capacity is taken by involvement with one and only one such possibility at a time. In Comedy there is no thinking past a given moment.

The evaluation of what makes a good story is itself a pleasant experience because it tasks one to recollect successful stories, i.e., stories that maintained interest and had satisfying resolutions. Several hundred years ago, William Shakespeare would take stories that were already popular and upgrade them in ways which garnered notoriety. He would begin a comedy with one or more innocent protagonists ready to choose a spouse. Their choices, however, were opposed by others who had no legitimate reason to do so and whose opposition would otherwise have succeeded but for the intervention of a benevolent and powerful agency.

The choice of sexual fulfillment as a motivating interest in the comedic myth is common for the simple reason that sexual maturity is a significant aspect of the claim of existence one makes. Children are entitled to food and shelter but the claims associated with sexual maturity involve responsibilities, commitments, and competition. In Comedy, parents, peers and rivals test this moment of transition—from childhood to adulthood, from dependence to independence, from self-effacement to self-assertion.

Comedy is not so much about the gratification of the desire but of self-assertion in a social setting which poses resistance. Societal pressure favors compliance—i.e., non-assertion, but moral success requires non-compliance. It is counter-intuitive, but we reserve a kind of moral loathing toward an individual who waits patiently for authoritative figures to choose their mate. Such an individual has given their life up for others to decide.

Typically, the comedic hero and heroine are in over their head, for they cannot, of their own will and limited resources, effectively circumvent organized social opposition. Happy endings are either the result of luck or the intervention of a benevolent power. This is why comedies are usually intoned with mirth and optimism, that is, because of a belief that there is a higher moral order of sorts that favors and rewards honest self-affirmation. If one knocks persistently at the door leading to adulthood, it is supposed to open even if one appears foolish in the process.

Yet there are comedies which end sadly. This is the case in Romeo and Juliet, as is frequently the case in life, and it is hard to say where the sadness lies. It might be due to a failure of a higher order overwhelmed by the opposition, or due to the fact that Romeo and Juliet gave up. Regardless of the ending, what is important is that one has identified with them, and wanted the world to open to them. Sometimes it does and sometimes it doesn't, but one knows what it ought to do and this "ought" is the moral of the story, its moral judgment.

The story tells us much about the way human beings "think" about moral issues—i.e., much more contextually and pragmatically than one might assert through a code of conduct. Moral values are revealed in narratives such as those reflected in the comic myth and literature. Now examine a list of related terms arising from one's engagement in Comedy.

1. Cognitive Science		2. Social Science		3. Moral Value	
Intellectual Processing	Intellectual Interest	Social Behavior	Social Interest	Moral Action	Moral Value
Declarative	Primary	Reward/ Punishment	Pleasure/ Pain	No killing	Existence

The terms arranged from left to right reflect increasing awareness of a value attached to social existence—from 1. a description of engagement with or commitment to the experience, to 2. a socio-psychological description of normative behavior, to 3. an experience of moral value embedded in myth. If someone were to say "I'm not sure I know what murder is, but I know it when I see it" they might be suggesting that the circumstances of a narrative need to be detailed and clarified before they can achieve moral resonance and certainty in such an evaluation.

Comedy is about the assertion of social existence, about asserting oneself within social confines as a living organism. Because the social parameters for such assertion is the subject matter of comedy, it is not uncommon to encounter a comic figure who is promiscuous, or who experiments with intoxicating substances. The social definition of "existence" may vary from place to place, but common to all societies is the placement of right upon existence. Taken in the context of comedic struggle, the terminology applicable to a moral baseline defined by Kohlberg or Maslow is understandably preoccupied with a human organism that thinks in terms of physical pleasure and pain.

But also in Comedy it is apparent that Kohlberg and Maslow, in order to establish a credible baseline, may have overstated the primitive. Given that social constraints, i.e., the corrective interference of parents, teachers, bosses, etc., are quite challenging, one's assertion of certain social rights and privileges, is a moment of inspiration. There is nothing at all primitive about it. An animal would cower before such resistance.

In Comedy we learn something about the timeless prohibition against murder—"you shall not kill." Is this command of Moses about preserving the biological viability of life? Animals are killed for food and social enemies are killed for protection. A conquered people are killed to make room for conquerors.

A right to existence is a social construct predicated upon membership, a requirement placed upon the collective to accommodate the existence of a member. Such existence is more than biological viability, but includes those features that make a life. One might kill, for example, by denying another a spouse, or by not respecting certain choices of which living consists. Slavery is a form of murder.

It is probably inaccurate to describe "existence" as a "need." In Comedy there is awareness that life is what is. A moral identity is the unique variant of an existence society embraces as common to all. In order for society to support life, its members must live, i.e., assert themselves as living beings. One does not ascribe value to other lives because he or she exists, but because they ascribe value to existence and regard the claim of viability in the form of choice as virtuous. Thus what appears to be proscription in the commands of Moses, i.e., "you shall not," is in fact descriptive of foundation, i.e., if you exist you "do not."

8. Romance follows Comedy.

Threats to existence—and its necessary features—must be opposed. In the course of such opposition, one encounters and condemns an adversary, and thus achieves focused and forceful motivation. One becomes a protector, there to advance a morality preoccupied with the elimination of evil.

In Comedy one learns that life consists of a number of activities, such as eating, working and childbearing, all of which make up a life process. Coming into the status of adult is a matter of claiming as a right the socially valid assertion of these activities. One who opposes this claim, or rather, who is unsettled and disturbed that others enjoy the exercise of liberties that society deems acceptable, is more than adverse to a given individual, but adverse to existence.

That such life related claims are common to all members of the social collective may escape the comic actor. The actor is too busy working his or her way into the group to notice that their individuality is enabled by the collective support and input of others. Group acceptance is a powerful moral event, accompanied by solemn oaths of commitment. But group resistance in comedy was only a test. Upon passing the test one experiences a moral commitment to the group strong enough to evoke a different moral perspective—embedded within a different story.

The romantic plot—often erroneously confused with a love story—is a quest fraught with deadly challenges. It follows a pattern described in the stage of intellectual development earlier referred to as "cumulative processing." The mythic romance is an adventure in which one's abilities are challenged, after which the challenges are eliminated from further consideration. The romantic myth is therefore processional in nature, a quest to encounter and eliminate obstacles along the way.

In such a quest great emphasis is placed on the confrontive presence of evil. The other side of evil is what we often refer to as "good," which does not have much in the way of content, other than to represent a socially based sanctuary where we build our homes and store food. Evil threatens this, and indeed in myth, preys upon it. Evil is not interested in joining our society, but takes what it wants without asking or giving in return.

For this to be a good story, evil constitutes a threat that challenges the full capacity of the hero. It is not a trivial nuisance, but a strong and active threat worthy of one's complete commitment. Nor is it a vague and ambiguous evil, about which one must weigh and subtly differentiate between various shades of value. The antagonists in a romance are frequently depicted as horrible and distorted creatures.

In this way the dualism noticeable in the world of the cumulative process—known versus unknown—assumes mythic importance in romance as light against dark, virtue against vice, peace against violence. Within such dramas we embed—or invoke the experience of—notions which convey value. This is again noticeable through the arrangement of such notions on a continuum—between the description of behaviors and the experience of value.

1. Cognitive Science		2. Social Science		3. Moral Value	
Intellectual Processing	Intellectual Interest	Social Behavior	Social Interest	Moral Action	Moral Value
Cumulative	Retention	Reciprocity	Security	No stealing	Property

These three groupings emphasize that mental processing, revealed in cognitive science, bears resemblance to group oriented behavioral motives and strategies. Something more happens as we move from the description of behavior (which includes motivation) to the experience of moral value in the acquisition and protection of social entitlements—as in the bestowal of good upon such creations.

Evil is likewise fashioned from insecurity that lives within relatively crowded social expectations. There is a right way and a wrong way to live together with others, the wrong way being a repudiation of the interests of others in the group. One may fear a personal enemy, but this does not translate directly into the moral abstraction "evil," any more than a gazelle's fear of a lion involves its moral condemnation of the lion.

The force exerted against evil is reciprocal, that is, usually similar to the force with which evil threatens. The threat of violence by criminals is countered by the infliction of violence against those identified as criminals. The threat of invasion by foreign sovereignties is handled by the invasion of those sovereignties. For that reason, the romantic hero is usually proficient in the very tools deployed by those identified with the threat.

Thus, to a social scientist, issues of good and evil are behavioral phenomena—i.e., merely what people think. But that tends to understate the force of the shared experience of moral value, or rather, what the "it" is that captivates the diverse interests of a social group. It lives in their minds and passes from person to person. A word we might use to denote this existence is "property," a society bonded by acquired privileges.

Property is the result of the assimilation and management of uncertainty. Through property we demonstrate control over our surroundings and establish rights which allow us to function strategically within very complex social environments. One achieves security—shelter, employment, access to food—by contributing value and engaging the social protection of one's acquisitions. Property thereby identifies a thing, a "protectable," and assigns the social blessing of protection to it.

Contemporary market economies seem to derive support from property as a moral value in the accumulation of wealth, clients, networks and resources. Commerce can hold the fascination of large numbers of people under a common romantic narrative, each living out an adventure of conquest and acquisition. The various licenses, memberships, equities and copyrights associated with the world of commerce become property only because great throngs of people embrace property as good and—almost by reflex—institute measures by which property rights are enforced.

The moral sanctification of property precedes moral decisions about which things should be protected and which should not. Linked to this value is the moral command "you shall not steal," which has clarity so long as one contemplates protectables. Moral clarity fades as one considers things which are arguably neither owned nor lost through the malice of others. A thing such as "freedom" is vague. To say "they want to take our freedom" could mean "they want to put us to work for them without paying" or mean "they want us to live in a more restricted and fearful environment" or mean "they want us to be unhappy." In the latter two instances it is more difficult to identify both the motive and the mechanism by which a given enemy can make such a transgression.[15]

The rule "you shall not steal" is, in effect, a maxim "if you exist, you are unable to steal without diminishing your existence." It is not a law of what should be, but a law of what "is" deriving from what one has already defined as those qualities necessary to a viable existence.

A liability of this perspective is that it tends to overstate its position, and attempts to dominate the entire moral landscape. Romantic assertions of good and evil are felt so strongly that they are often confused with religious experience. There is no alternative except to be "for us" or "against us." Other moral viewpoints which demand more nuanced evaluation, which would handle a violent enemy without violence are regarded dismissively, or condemned as weak.

9. And so there is a moral purpose for tragedy.

Is moral development at an end after evil is defeated? Some might think so, but there are others who are unsettled at the thought of a life of complete tranquility. A great warrior who has laid villains all to rest, who no others dare challenge, and who has claimed his reward, may yet encounter a threat which cannot be slain.

To define selfhood in terms of a set of evils that one is not, while moralistic, is morally unsatisfying. It succeeds to a degree, and then fails. The story of this failure is tragedy. People live through tragedies because they wish to know who they are, i.e., because the moral foundation of being a happy person, is being a person. This is unachievable through direct engagement, and involves a kind of uncertainty unacceptable to a romantic hero. Tragedy involves an evil which does not allow itself to be confronted, but carries sufficient potency to humble the greatest of us.

The mythic tragedy follows the pattern of the stage of intellectual development earlier referred to as "serial processing." The tragic actor reaches across to the unknown and includes it in their own experience without knowing it directly. The unknown, a certain evil, is embraced by consciousness and included within it without being defeated. The unhappiness is manifest in the fact that evil seems to win, but not without being redefined. In tragedy the hero is figuratively undone and dissolved, removed from their mental castle and faced with the prospect of being what they hate. Less attention is given to the active opposition to external threats, and more on inner turmoil associated moral ambiguity.

Strangely enough, through the apparent victory of evil, evil is redefined in a way that allows good and evil to cohabitate. One accepts the other. It is therefore appropriate that the word "intimate" means "imply," the defining characteristic in the logic of demonstration, while also denoting a sense of closeness to and belonging with another person. To intimate is to ascribe value to another experience deemed exterior to one's own. To be intimate is to exist in a state of connectedness to the experience of another, and to value that experience even as it opposes their own interest.

The word "tragic" is thus a bit misleading, because embedded within all the pain, something quite good is happening to the hero. Though there is certainly a cognitive feature to the growth that is occurring, the significance of that is in a sense trivial compared to the emotive changes occurring in the emergence of a social identity. As we acquire this need for intimacy, we practice altruism—doing good to others for their sake— and adopt moral values that renounce selfishness and affirm trust.

1. Cognitive Science		2. Social Science		3. Moral Value	
Intellectual Processing	Intellectual Interest	Social Behavior	Social Interest	Moral Action	Moral Value
Serial	Implication	Altruism	Intimacy	No Adultery	Trust

This compilation of notions implies a world is fertile with tragic potentialities, for little if not all of what is defined "evil" is more ambiguous than what one may in a mythically controlled romantic world separate from "good." The experience of value rises as one resolves their fear and anxiety over the sense of evil which is unknown or uncontrollable. This is the successful conclusion of tragedy, which transforms various estrangements that isolate us into part of a newly defined good.

To the tragic actor, therefore, the sacrificial act of a zealot is nothing special. Genuine altruism is less about sacrifice than it is about acceptance. It is not about displacing the significance of one's own life, but about redefining self in such a way that the value of other falls quietly and naturally into place.

Tragedy plays out in dramas that are much less spectacular but significantly more disturbing than one will find in the various action genres of the romantic myth. Our tragedies are subtle and selective in subjects they address. Evil has lost its adversary presence and tends to appear as a duplicitous betrayal from within one's own house. Such was the case for Hamlet, who faced betrayal by his mother, for Othello, who faced the alleged betrayal of his wife, and for Lear, who faced the betrayal of his children. For each the sanctuary of home was broken.[16]

It was not the betrayal, however, which destroyed them, but their reaction to it. Though perceived as a struggle against betrayal, it was more of a struggle against the estrangement arising from a sense of shattered entitlement. In order to intensify interest in the tragic hero's inner struggle, the so called "evil" with which he or she struggles may include the intervention of a deceiver, in order to show that the evil that agitates the tragic hero was uncertain and illusory. The audience wants them to survive, to rise above it, but knows they would have to change in order to do so. The self-image formed out of a simplistic division between good and evil had to dissolve, suffer destruction, to remake an identity that could accommodate betrayal.

The tragic hero may fight, but it is not the fight of a warrior against an assailant, but more like self-inflicted suffering intoned with suicidal angst. The decision to fight is more of a disquieting acquiescence to the inevitable. That was the way Hamlet wanted it, as he opposed a claim of marriage and familial bond posed by his uncle and supported by his mother. His actions were aimed at exposing the adulteration which had occurred, i.e., simple comfort feigning intimacy, even if he would die in the process.

Lear and Othello were similarly preoccupied with the issues of feigned commitment and trust. Perceived betrayals were an invitation to ponder a distinction between familial affiliation based on changeable personal interest, where individuals collect together for their common benefit, and that of family based on intimate mutual commitment and validation. The task before them was to differentiate between false, or adulterated intimacy, and that which is genuine.

In more common lives it could be part of a crisis in mid-life, as one spouse abandons or betrays the other because the familial bond fails to elevate beyond a contract of exchange and each experiences an irrecoverable sense of isolation. Marriage either reforms or the partners seek intimacy elsewhere. Any number of experiences qualify as tragedies, and it is difficult to say who among the public speaks with authority about them. One might search among the occupations that offer assistance in the reconstruction of shattered egos.

At some point after the worst is over, one may reach a point where the value that fueled the experience stands out with sufficient clarity to be named. This value, however, is not the sort of thing that one possesses. "Trust" is an enabling value, a bridge between one existence and another. A value such as this, which successfully links two persons, is a breakthrough. One emerges from the lonely sensation of being surrounded by many, but close to none. It could be the first moral experience one consciously associates with liberation.

Trust is a state of comfort, not with the fact that others will do harm, but with the fact that they might. The other on whom one depends is in a position to, and often does, hurt one the most. Yet they are not evil. They are merely "other." It is a challenge, such as it is, to forgive them as one forgives oneself. Trust is a calculated risk that labors over the need to cede control. Being trusted, as in receiving trust, requires very little effort. The more challenging and dangerous act is to extend trust to another, which is the whole point of tragedy, i.e., to trust wisely and meaningfully.

The word "adulterate" means to create or support a false and inferior version of something, while the word "adultery" identifies a sin against the presumed intimacy of marriage. To commit adultery is to adulterate trust, or rather, to receive trust from another while not participating in it. It is a different kind of sin than the sin of murder, which addresses the value of life, and the sin of theft, which addresses the value of property. Adultery is a failure to honor and support intimacy.

In betrayal, one mocks and belittles intimacy and thereafter finds oneself alone despite the generous affections of others. Adultery consists in temporary or insincere commitments of many kinds and in the transgression against any genuine relation of trust. A businessman who betrays the trust of his partner commits adultery, as does a lawyer who betrays a client.

One does not therefore attempt trust without valuing it. That is the resolution of tragedy, i.e., a decision to accept the risk motivated by the knowledge that one is bonded to "other" not so much by contract, but by affection. We do not trust because we need to receive affection, but trust because we need to experience other with intimacy. We give trust because trust is what we are.

10. Having survived tragedy, we then become impressed with irony.

Some that have experienced numerous integrations into unfamiliar groups or societies are apt to be impressed with the constants that seem to govern such events. There is something universal about friendship, for example, which affects human beings as they move from one social group to another, and which we associate with something valuable.

The experience of social immersion—together with the moral attachments of loyalty, sacrifice, altruism, etc—precedes the emergence of a new mythic experience, one which permits and encourages detachment, and moves toward the discernment of organizing social principles.

The soul's acquisition of a value that accommodates and surpasses lesser commitments is the topic of the mythical genre of Irony. Irony follows the pattern identified in the intellectual stage of development referred to as "parallel processing," the pattern of deconstruction and reconstruction. This is the logic of order and understanding.

Satire is a form of Irony because it belittles things we—or those having authority over others—hold dear. The belittlement usually consists in the comparison self-important people to events and circumstances which are, ironically, simpler and more powerful than they are. Some ironies are called "histories," which are good at making the point that historic events—labeled as such because their great effects—are controlled by a few people preoccupied with simple problems.

There are more contemporary ironies, as in the domination of a criminal or military underworld in politics, or the political victory of a single person who manages to take an effective or persuasive stand against better equipped and funded opponents. Irony portrays an individual aligned with real power vanquishing others who think they are powerful, but are mistaken. Moses' parting of the sea was irony. Jesus' resurrection and the ensuing formation of a Roman Church was irony. Henry V's victory over the French legions and the defeat of the Soviet Union by an Islamic resistance in Afghanistan was ironic. History is full of ironies.

The main character of this story is typically cast in a predicament of having to choose between being comfortable with meaningful but limited projects and taking on great tasks. Before Achilles left Greece he was told that he could remain in happy comfort, or die famously in Troy. Divine intervention in the Iliad, as in the stories of Moses, Mohammed and Jesus is told (and remembered) in large part because the human soul is unsatisfied with limited tribal affiliations and seeks social universals that validate a moral order.

In a manner of speaking, the "other" of tragedy is abstracted into societal "other" governed by principles which connect the human to deity. It is one thing to have friends, and to be close to them, and another to believe in friendship, or rather, to understand that friendship is a quality in life that permeates a society of individuals, and in which all may participate.

Once again we find a number of related ideas embedded in narrative, which, depending on emphasis, go by different names.

1. Cognitive Science		2. Social Science		3. Moral Value	
Intellectual Processing	Intellectual Interest	Social Behavior	Social Interest	Moral Behavior	Moral Interest
Parallel	Order	Law	Status	No Perjury	Honesty

Graceful interaction with people is a function of maintaining sensitivity to their unique perspectives, and allowing them to be who they are. In irony, one is apt to become a social theorist and ponder the order of social cohesion and solidarity—written for the most part into a system of law. There is no discrete point of termination to this activity, except that there appears to be, as described below, a point at which one achieves satisfaction.

Maslow's interest in "status" is only part of the ironic experience. It may not be, as Maslow once argued—a need, because a perceived need for status is, by its nature, unstable. To think of a need in terms of the socially deferential behavior of others misses the point. Although irony involves one in the examination of self-importance, it is something designed to tempt rather than satisfy the ironic protagonist. Self-importance is the adversary. The ironic heroes are lured by vanity and pride into a situation from which they must extricate themselves. If he or she fails, then they have succumbed to something loathsome.

This is an idea that bears repeating, because a craving for social status can be irrepressible, and cannot be satisfied until it is displaced by a value which resembles status but is not the same. Instead of imagining status as a property that one possesses to the exclusion of many others, the significance of a human life is affirmed as something shared by all. For this reason, irony resolves in part by giving "status" a different name—perhaps "dignity" or "respect."

That a craving for self-importance leads one to humility is itself ironic. Imagine a story of one who organizes a society of deferential personalities, only to be cursed with the alienation and exile derived from the insincere or unworthy admiration. The antidote, also available in the ironic story, is to embrace "dignity" as a social universal, and serve as donor rather than consumer of it. In this way, one acquires respect as a value insofar as they give respect to others.

A grasp of this idea illuminates the ninth command of Moses: "you shall not bear false witness." One can by lying cause another to lose life, property or family, but such lying is in fact only another version of killing, stealing and adultery. "Bearing witness" is not just about what one says of another, but about how one holds another in thought. Unique to this command is a requirement that one honor the "dignity" of others, of giving "respect."

This is the only moral precept of Moses which holds one responsible to truth and is, therefore, as deep as our understanding of what we are. At one level it is a command against falsely injuring another's reputation. At another level, it encourages one to be sensitive to the fact that the significance of another may be different than a sin one commits on a given day. To publish scandal about a given person who in a larger picture lives life righteously and honorably, in a sense "bears false witness" because the message bears no relation to the significance of the person.

At the highest level this command encourages one to affirm the intrinsic goodness in another who by all standards has lived a life of wickedness, i.e., to regard such wickedness as a lie about the "true" person. One who genuinely understands respect and respectability responds to immoral behavior with bewilderment and compassion. Yet such understanding, i.e., to achieve a moral perspective of such depth, requires a highly evolved spirituality.

In acknowledging this, however, we get ahead of ourselves, except in the simple recognition that what we think of as "spiritual" often bears an ironic relationship to what we think of as "moral." For a number of reasons—which shall become clearer from further examination of freedom—a highly evolved spirituality overreaches the achievement of moral satisfaction. Moral principles are practical measures that help to define us as social beings. With that in mind, let us admit it is enough to be honest, that is, to have respect for the truth, and to be stubbornly mindful of the disparity between what we think we know and what is.

11. The passage through irony involved us in the last mythic narratives in which moral value is found. There is a return to comedy, but social scientists—here Maslow and Kohlberg—are convinced of a fifth stage, and speak theoretically of a sixth.

1. Cognitive Science		2. Social Science		3. Moral Value	
Intellectual Processing	Intellectual Interest	Social Behavior	Social Interest	Moral Action	Moral Value
Declarative	New Primary	Justice	Self Actualization	No Coveting	Autonomy

Maslow suggests that progression to a fifth tier of development places one in a state of contentment, or rather, completeness of self, which he calls "self-actualization." Logically, the idea suggests a termination to moral development, for what can be more complete than complete? Moreover, there is a certain moral grace to the achievement of completeness; in that one's moral longing can be satisfied.

The same can be said of *justice,* which in the Socratic moral universe represented complete agreement (harmony) between man and society. Unlike an ever growing intellect, moral development is context bound and therefore attainable. Finding agreement between one's self and their society is by its nature elusive, but resonates with greater depth than an intellectual achievement.

We must be careful, therefore, as psychologists or philosophers, not to attempt to wring out of moral development a solution to the human predicament. It would be, in a sense, like physicians, bolstered in confidence by the discovery of antibiotics, using the same to attack diseases which have nothing to do with bacteria. As with the overuse of antibiotics here one might, through the overuse of morality, diminish it.

That being said, let us try to imagine another order of morality. The progress reflected in Maslow's "actualized" self, is the successful resolution of the ironic myth. There is no place to go except a return to comedy. Such a return would not be the comedy of an individual making a claim of membership within a society, but the comedy of a society—or representative thereof—seeking membership among other societies. In intellectual development, a new order regards the old order as a single object within a larger field.

But moral development is not about developing the wholeness of an object, but about wholeness of self, and the self cannot become an object. Likewise, though we may identify with a society, we are not and cannot be one and the same with it. When we have moral completeness (wholeness) our moral development is concluded. We may contemplate ethical values of a different nature, but they do not entail moral struggle and mythic resolution. We can and do consider other kinds of freedom—as in the value of political and spiritual freedom—but we cannot do so effectively without first appreciating the limits of moral experience.

Moral freedom is much more challenging emotionally than intellectually, and for that reason is coded into narrative. In this way, among others, moral freedom does not consist of choosing obedience to reason, but in choosing to include the interests of other human beings in the formation of one's own interests. The remarkable synthesis that emerges is present and unmediated, and released to experience as a story unfolds.

One may deconstruct the identity that comes out, but the process is likely arduous and inconclusive, just as the construct of self is arduous and intuitive. One cannot make an object of their identity, but they can in a different sense, arrive at a state of sufficient completion that they cease to labor over it. This state, which to others may appear humbled is better described as "quieted," as one ceases the mythic dialogue of self-justification.

As one attempts—through the application of intellect—to deduce moral precepts from an ultimate good, they become distanced from that good, and may fail in the moral experience of selfhood. The inference of good as a concept related to other concepts is not a moral affirmation. It is not choice (personal identification) but the abdication of choice to other.

The fifth moral command of Moses, i.e., "you shall not covet" is a command that reinforces the idea that moral laws are designed to serve rather than displace the formation of identity. It commands one "be content with who you are." It is, in effect, a repetition of the first moral command "you shall not kill." One cannot value existence as an irreducible moral good while measuring that good against the existence of others. Coveting is a form of suicide, a way of handing one's existence over to other, thereby refusing the comic challenge.

The myth of Irony marks a final point of moral arrival. It resolves a socially dependent and covetous desire for acceptance and approval, into an independent and autonomous sense of dignity derived from a trans-personal principle of self-affirmation. The possession of this completes and satisfies moral longing. Mythic irony delivers one to a self-affirming and generous attitude to which, morally speaking, there is little to add.

These principles merely summarize that toward which moral freedom was evolving. There may be a higher good, but whether it is right to define it in moral terms is debatable. If there is an ethical stage which follows this one, is it appropriate to think of it in moral terms?

The question is, perhaps, premature, because the only other type of "ethical" existence we have mentioned is the "value" intrinsic to intellectual functioning. What we have been finding in the moral experience are values that grow and function as animating motivations, and have been diversified into types triggered by certain narrative events. An attempt to abstract a unifying principle misses an important lesson about moral value, i.e., that moral values collect around and serve as extensions of an experience of ego—a compound experience resistant to tidy summary.

Thus, the articulation of a highest moral principle should be made cautiously as they are usually projections of perspectives that are coarse and limiting by comparison to the richness of the human experience. Utopist imaginings tend toward such deficiency—they seek to extract more from moral experience than it can offer, and no one alive is perfect enough to administer them.

In fact, the erection of a moral pyramid designed to reflect and control progressively larger social domains—organized within intellect—presents a moral paradox, of sorts. While the intellect is involved in expanding the field of objects available to consider, moral value is released from that imperative, and through narrative tailors its universe to facts which serve in the experience of value. It is not trying to get larger, but to achieve satisfaction.

The subject material is not the social group but the moral autonomy—freedom—indicated by an agreement between one's personal existence and social existence. Surrender to the group either through self-denial or through various subtle intellectual fascinations with social norms, are ways of denying one's autonomy, of coveting. The "me" becomes lost in "we."

Societies are not autonomous moral entities, but rather, the surroundings in which moral identities emerge. To speak of the "moral interest" of a social group is problematic, and it is similarly problematic to refer to the political dialogue between social groups as a moral dialogue. They may support different kinds of moral identities, and thereafter attempt to dominate each other without understanding that their differences do not have a moral solution.

More tragically, the leader who becomes highly identified with the social group and its unique moral perspective may dissociate from their existence. Their isolation may bring about flailing attempts to restore a sense of connectedness to their humanity. They may commit immoral acts only to rehearse feelings that bring them back into moral focus.

Moral freedom thus recommends respect for both the significance and the limitation of moral self-affirmation. One might at some point disengage from the application of moral rules toward the resolution of disputes in which such rules are the cause rather than the effect of social discord. At this point one, quite significantly, retains a moral identity, while resituating to an ethical posture that is quite different than moral self-affirmation.

12. This process of reorientation is possible only when one admits that they must—in a way—change the ethical universe with which they are preoccupied. This does not occur because one moves to and from an epistemic and ethical reality, or in and out of valuation. If we were to take a lesson from our examination of intellectual freedom, it would be that valuation—an ethically positive assertion—is essential to cognition.

The question we address in an open discourse on "morality" is how to make an ethically positive assertion about our social identity. Important to this examination is the recognition that ego and identity are far less tractable than most objects we encounter as part of an external environment, and do not adhere directly to our nascent cognitive mastery of that environment.

A correspondence between the interests driving the logic through which we construct various existences in our surroundings, and cherished moral principles which have been preserved in our religious traditions, suggests that "existence" permeates consciousness as an organizing principle. It is no great leap of faith to see an invisible hand at work and call this hand "genetic." Whether there is a way to take this correspondence in a different direction, and construct a theology—of sorts—from it is beyond the scope of this examination.

One must also respect the religious implications of a moral value structure that supports "existence" both in the abstract, and in its application to the formation of human selfhood. Indeed, the fact that there is alignment between information processing modalities and the mythical patterns which support moral values canonized in Judaic texts is fertile ground for the advancement of the monistic precepts of that religious tradition.

But before moving too far in that direction, it might be useful to examine a religious tradition often considered a neighbor to the monism of the various Abrahamic faiths—Judaism, Islam, and Christianity. The Buddhist spiritual tradition promises enlightenment—and an elevated and comprehensive state of happiness, as a consequence of meditative practices and adherence to moral rules. It does so without being preoccupied with theological statements about a God. The moral rules organize into "five" which bear strong resemblance to the five moral commands of Moses, as indicated below.

Judaic Commands	Buddhist Precepts
(For all Persons)	(For Laypersons)
You shall not	You should refrain from
Kill	Killing
Steal	Stealing
Commit adultery	Sensual misconduct
Bear false witness	Lying
Covet	Intoxication

The resemblance is so strong in the first four commands that one is tempted to resolve dissonance presented by apparent dissimilarity between coveting and intoxication. Such resolution is quite simple, actually, given our discussion above. The Buddhist's advice against intoxication derives from the value of "mindfulness" or rather, a calm awareness of one's own consciousness. Intoxication, much like coveting, is a way of surrendering one's own being to other. It is therefore not surprising that a religious tradition emphasizing a meditative rather than a moral strategy to the achievement of religious consciousness might be more concerned with the effect of intoxication on one's existence. But in both instances one's interest in autonomous existence governs the formation of a behavioral precept that tends to preserve moral autonomy.

Given enough time and motivation, one might develop an interesting theory of correspondence between the two sins of coveting and intoxication. It seems that the depressing sensations of unworthiness and low self-esteem attract the use of intoxicants to medicate the pain wrought by a dis-autonomous and inauthentic ego. Psycho-therapists frequently remark on how drug abuse ultimately reinforces sensations of this kind.

Whether we are talking about a complex of ideas in one universe of depression, worthiness and failure or another universe of addiction, substance abuse, and self-control, we are in both instances offering a summary, of sorts, of a pathology that might flow from essential moral failure in establishing an effective identity.

Contemporary psycho-pathology has observed that pain inducing or maladaptive egos form within consciousness with regularity, and that these egos are not properly regarded as a "true" or "healthy" expression of consciousness. They are, nonetheless, self-organizing, put up a fight when challenged, and can manifest in a remarkable array of physical symptoms. Psychologists seem to agree, however, that the treatment of maladaptive ego does not require subscription to a monist theology, and often promote the Buddhist strategy of disengagement from such ego through meditation or discursive analysis which allows destructive ego constructs to be reformed.

There is, in fact, a fair amount of literature which suggests that ego is completely dispensable, and that we are better off without it. While it is difficult in evaluating this literature to be certain that terms are being used similarly there as they are here, a very strong consumer interest in the mythic narratives which reveal "self" and "ego" suggests the existence of authentic selfhood is an important aspect of personal fulfillment and contentment. The elimination of "ego" is problematic, though a given sense of ego can prove unreal or unworthy of belief.

Having a list of irreducible values necessary to the formation of selfhood and understanding how they relate to such formation, can be useful in the differentiation of real and false egos, and in the diagnosis and treatment of suffering. As in all things, the assessment and acceptance of variance between persons benefits from the understanding of constants. That we seem to have uncovered moral constancy in motives which mirror the interest(s) driving cognitive development can be helpful in sorting through and interpreting variance of the many different forms of moral expression occurring within those constants.

We are, in other words, likely to see many different ways of coding moral rules, though we will likely be grouping this diversification within categories which affirm life, property, trust, respect and autonomy. Though diverse in its narratives, authentic selfhood implies cohesion among these values.

Peter Gibson Friesen

4.

Political Freedom

The creatures have only one sense: touch. They have weak powers of telepathy. The messages they are capable of transmitting and receiving are almost as monotonous as the song of Mercury. They have only two possible messages. The first is an automatic response to the second, and the second is an automatic response to the first. The first is, "Here I am, here I am, here I am." The second is, "So glad you are, so glad you are, so glad you are."

Vonnegut

1. As the preceding chapter revealed, moral behavior is both more and less remarkable than many moralists would argue. Moral experience and the formation of a socially based ego is primevally integrated into our psyche and, like the acquisition of a language, places us within a society of others who share our values.

While it is important to acknowledge the role of moral value in the formation of cohesive social systems, it is—as indicated in the last chapter—important to acknowledge its limitations. Human behavior is marked by both assembly and disassembly. Such propensities are most visible in modern political systems.

As suggested earlier, the job of forming and managing social organizations is easier within groups—either small or large—bound by a shared sense of right and wrong. There is a point, however, in the growth of any given society where one group bound to its own moral perspective separates. At this point the issue likely to dominate one tasked with preserving the organization is how to deal with separation. Does one become more forceful in the promotion of its perspective, or are there ways to accommodate the diversification of moral value? The choices one makes at this point are apt to have a significant impact on the kind of organization that results—i.e., permissive or authoritarian, large or small, uniform or diverse, peaceful or rancorous.

How does one establish and maintain order within a social environment that is in many respects uncontrollable? In order to understand the navigation of these situations, it is necessary to examine the way people set priorities within systems of communication. This will require some explanation.

Throughout the world there are many societies, which occasionally coalesce into nations, and where there is less than complete agreement over what range of moral values are acceptable. Ironically, the very normative systems crucial to the assembly of nurturing, protective and stable societies are often the source of conflict between societies. The mythic power of moral value—intense and unmediated—struggles when tasked to comprehend and accommodate other views.

Such accommodation tends to occur in a process by which one society communicates to another. There are a number of different ways to communicate. At the most direct and immediate level, there is the medium of spoken communication. At another level, but one which is more traditionally associated with politics, is the medium of law.

There is hardly an animal which fails to communicate with others in its species, but the communication we refer to as "language" is often used to set human beings apart from other species. Our communication is full of approval and disapproval of each other and of agreement and disagreement over what ought to be. All of this may be crudely assembled under the rubric of "politics"—action designed to influence what others find acceptable.

There is good reason to exchange influence—one identity to another—as effective human societies must organize and select constituencies, leaders, and institutions. They must make and administer laws. It is obvious, at one level, that human beings do this because it is in their collective best interest to do so. But this hardly explains why political formations vary in their stability, or why oppressive regimes frequently prevail over democratically conceived governments.

Since divergent views of a viable and decent government are resolved through adaptive and competitive political caucus, the task of establishing a just government bears resemblance to a game. The use of a "game" as an explanatory metaphor is receptive to the diversity and interdependency of perspectives that participate in it. It is a way of dislodging attachment to the moral viewpoint of one person or group of persons, and becoming refocused on the patterns and processes that support the continuity of the social system. It encourages the participants to place more value on the continuity of the process than they would on any particular outcome.

The maintenance of a peaceful and continuous relationship with government is important enough to most rational persons that they would tolerate some injustice in order to preserve that relationship. This is due in part to our intuitive awareness that there is less than perfect consensus over what is just. Indeed, social leaders may gather together at times to propose adjustments to the processes of government, but when they confer to discuss what kind of government is optimal, they soon learn that what is "just" to some is oppressive to others.

The easiest way to resolve the political enmity stemming from such a conference is for the groups to form separate kingdoms. This is difficult to accomplish peacefully, especially in a crowded modern world. As the need to establish government within larger geographies and more diverse cultures increases, so too is the need to establish a technology of government based on a more inclusionary credo.

The challenge derives from the strength of commitment a given moral code evokes from its adherents. This was a matter discussed with "moral freedom." Implicit in such a code, therefore, is a limited range of acceptable behaviors. Since the viability of one's individuality depends in significant part on the capacity of others to recognize it as legitimate, there are strong intuitive barriers to certain kinds of experimentation.

Politics deals with the problem associated with the proliferation of diverse societies, each representing a moral platform, which, absent politics, could not be reconciled. Politics addresses the fact that diverse moral systems based on diverse identities are prone to conflict, and that this is because the human being reposes moral value within a part of consciousness which is not easily accessed by intellect. Most often the reverse is the case, i.e., moral attachments drive choices that precede rationalization.

If there were but one identity which could be posed as the moral solution to interests in conflict, there would be very little need for politics. The ethical universe is limited such that the only choice is whether to comply. By forming such a society we might limit the need for politics, there being a preexisting agreement on how one should behave. In such a universe, the only behavioral questions are factual ones.

There is nothing political in the behavior of a mound of ants because there is no disagreement among the ants over what is right—no dispute over what makes an ant what and who it is, and what it ought to be. The central feature of politics is disagreement between different moral codes. Political disagreement is most noticeably manifest in a conversation between two "moral" individuals each claiming moral entitlement to the same thing, and unable, after agreeing on what the facts are, to resolve their differences.

Even so called "totalitarian" regimes acknowledge the importance of persuasion in the affairs of government, and that the task of governing is eased by a cooperative citizenary. Autocratic governments tend to appropriate the persuasive media in which social ideals are clarified. In past eras, these were churches and other religious organizations. In modern representative democracies, religious partnership with government is prohibited.

This step, i.e., the deliberate protection of religious movements from the influence or interference of governmental agencies, was part of a growing recognition that religious ideologues do not govern effectively in culturally heterogeneous societies. When churches, representing distinct and conflicting moral identities decide to abandon persuasion, their recourse, often expressed moralistically, is war. War is a failure of politics.

The avoidance of war requires sufficient popular detachment from a single moral identity to allow peaceful communication between moral systems. There are personal as well as social benefits that accompany the diversification of moral perspectives and the formation of governments which accommodate them. As a culture—or civilization—we act as if we know it, but we do not always act as if we understand.

2. We have in our discussion of conscious experience been shifting inward—from "object" to "interest" to "identity." In order to understand politics and political freedom, we must move inward another step to a point of reference which will be called "deity"—i.e., something worshipped.

If we were to encounter a species that bore no physical resemblance to the homo sapiens of Earth, we may yet feel close to that species upon the discovery that the members of that species worshipped. Though quite a few human beings claim to worship nothing, i.e., to have no deities, their position is attributable to a different understanding of what the term "worship" means. What they mean is they do not engage in the active recognition of an objectified deific presence, or otherwise break into overt religious observances.

Worship is a state of surrender to some authority presumed to animate existence and the treatment of that authority within a game of language as certain. The scientist who renounces worship as "unscientific" is, in the process of renunciation, worshipping, i.e., yielding not to a conceptual object, but a presence, which includes, among other things, negative sentiments toward those who have imaginatively supposed the existence of deific beings. He or she seem to know what makes things happen, i.e., what the necessary properties of existence are.

The renunciation of an object deity, sometimes referred to as "atheism," is a belief animated by a deity that so completely dominates one's subjective landscape that it cannot receive a face. One form of worship likely to result in the rejection of object deities of all kinds has already received much attention in this book. It was the "reason" which produces the languages that other animals do not seem to have, and that makes life on earth comfortable. Thus, the atheism of reason may be less a matter of the renunciation of worship than the worship of a deity that is so jealous of competing deities, that it will not acknowledge them.

Examine the behavior of a single individual in two different ethical worlds, a change which commonly occurs between different athletic contests. A golfer, for example, acts under the dominion of a genteel credo which emphasizes and rewards precision, the humble acceptance of misfortune, and the meticulous observation of rules. The thought of carrying this set of ideals to the game of hockey, however, would be preposterous. The games are, in a sense, animated by different deities.

The use of a "game," however, to represent a state of ethical orientation only partially conveys what a deity is. A game is something with which one can choose to engage or disengage, but a true deity is never really disengaged, and follows one in and out of sport. A wealthy person may, for example, join a corps of volunteers and spend a year living with the poor in dire circumstances. Though they may have bonded with that life, they will likely not have acquired its deity, because they will not have assumed the burden of becoming irrecoverably poor.

A deity is a confining experience. It holds identity within limits, establishing the range in which an identity can achieve physical, moral and economic acceptability. Since survival strategies almost always include social groups, deity is almost always a shared social experience. We don't control it, but it does intervene in our affairs, and, in fact is present in everything we say or do.

A fair assumption, then, is that the content of one's deity, though known only through its effects, is bound to the circumstances of one's life. The compound ideal that a deity introduces to one's experience sanctifies what we are. A deity addresses us where we are, saying what is special about being human. In that sense, the presence of deity is an ethical presence because it influences one to conform to what one ought to be.

It is, in a sense, pointless to attempt to articulate one's own deity. The usefulness of defining a deity is no less elusive than an attempt to define a "reality." As noted in a preceding chapter on "scientific freedom," the use of the term "reality" is what one supposes is the cause of what one senses, and helps to anticipate further experience. If one did not respect such influence, it would be difficult to navigate a given reality effectively. Similarly, "deity" represents an assumption we make about the existence of an ethical authority beyond our control and is sensed as independent because it seems to intervene in what we are doing.

This authority is sometimes called "conscience," a word which tends to limit its authoritative presence to single individuals. It is also referred to as "God," a word designed to emphasize a presence that is experienced by many. By naming and defining a socially embraced "God" one prepares to experience the authoritative assertions of conscience, in much the same way one anticipates perceptions by developing a theory of reality. In both instances one enables oneself, as an organism, to respond to the influences from which consciousness is made.

A skeptic may say that to call such an influence by name, such as to say "God," overstates permissible inferences that can be drawn from the experience. But "overstatement" is a relative term. To refer to the attractive tendency of celestial objects as gravitational "force" is an overstatement, especially among scientists who continue to debate over what gravity is. It nonetheless helps to use the term "gravity" to engage thought in the anticipation of events.

The proper objection to the use of the term "God" is that it fails to distinguish between what spiritually minded people think of as an infinite creative force and the status of one's experience of that force at a given moment in one's development. Assuming the existence of an ethical force that exceeds any sense of boundary we can imagine, the experience of what we objectify as "God" is more of a placeholder for something else, and thus to name the deity established to reflect an infinite God is presumptuous and arrogant. It tends to limit the development of an ethical presence in one's life in the same way that a dogma can limit one's receptivity to evidence of reality.

This insight is useful in matters of governance and in the formation of social institutions. The simple recognition that what one calls "God" is a special adaptation to a boundless creative presence encourages government to accommodate the numerous deities apt to form in a heterogeneous mix of human societies.

Since deities tend to stabilize social systems, it is not uncommon, as suggested earlier, for government to claim an authoritative relationship with God. This claim includes, among other things, divine authorization for the government, and conscious devaluation of the authority of personal conscience. This tendency in the formation of social systems encounters problems, however, as moral perspectives within a society diversify.

The problem is not solved by simply becoming more ethically permissive. With nothing taken seriously, political conversation has difficulty with getting focused on the negotiation of real controversies. The experience of politics moves toward a hyper-fascination with the personalities and the strategies associated with the achievement of victory in the political process and avoids the exchange of perspectives necessary to achieve peace. By contrast, political freedom involves the diversification of moral viewpoints within an elevated ethical consciousness.

3. We have, on more than one occasion, emphasized that mind combines diverse experiences into singularities. The experience of objects is different than that of the interests which motivate their discernment—though they work together. The same is true of interest and identity. These experiences are private in the sense that they exist within the mind of one person existing separate and apart from another. But language, it seems, provides us with a way of sharing the whole of this experience with someone else.

One may not have an urgent need to know that certain intellectual constructs are acceptable to others, such as ones perception of a tree. Yet the failure of another to verify that it is there might arouse concern over whether that tree is real or imaginary. For other more complex constructs such as the composition of an institution or a theory of Justice, validation is more important. For experiences we refer to as "interest" or "self," validation is indispensable and formative.

In saying this one may become too focused on spoken language, and exclude other media of communication that involve a more direct way of making an impression on others, as in drawing pictures. A dolphin, for example, that is accustomed to the fabrication of vivid and precise images from sonic reflections in water, may well have developed a way of communicating with pictures that require more (or less) conscious attention to the identity of the communicator.

Our sense of comfort with who we are involves much more than what we are willing to say. With our various inflections, and in the way we organize what we say, most human beings seem to know intuitively that who they are is very difficult to conceal and that what they intend to say is not apt to be heard by others who dislike them.

The same is true of a government operating through a system of laws. Political freedom examines the communication between sovereign and subject, which occurs through a legal medium. In the case of totalitarian rule, communication is one sided and intolerant. Under a freer government, politics is more like a conversation, a blending of views.

Common among existing models of spoken language is the division of language into two tiers, one that addresses the meaning of words, and another that addresses the structures into which those words are organized. The former is called "semantics" and the latter "syntax." Law is similarly organized into a two tiered model, dividing law into "substantive" and "procedural" laws, one that permits or proscribes specific behaviors, and another that establishes rules by which behavioral standards may be applied to the formation of judgments.

We have already examined stratification in developmental sequences—of intellect and moral value—and so an interest here in the stratification of language would be expected. Can language be explained with two levels, and if not, does our experience with the stratification of human development provide us with assistance in identifying a more complete order to language? Perhaps, but implicit in the hierarchical arrangement of ideas is a metaphysical proposition—of the contour and terrain of the reality we experience.

In spoken language, a science of syntax, for example, might yield a set of syntactic rules. The rules then appear to represent an axiomatic system that defines "reason." In law, a science of procedural rules might yield a set of rules that define "justice." These are further attempts to have objects bestow value on objects, and to place intellect (the maker of objects) on top of a metaphysical pyramid.

Indeed, the academic world has poured much energy into the debate over the susceptibility of speech and law to structural reductions—e.g. "structuralism" in the spoken language and "legal positivism" in jurisprudence. Others, some of them referring to themselves as "deconstructionists" argued that "meaning" in language is not objective, but dominated by values and intentions unique to each speaker and listener. As an imperfect instrumentality, interpretations of language are by their nature matters of approximation and social negotiation.

We can become highly focused on the structural aspects of law, but as we attempt to understand it, the intricacy of interests revealed (deconstructed) in its rules tends to mock the notion that a value neutral position is plausible. Conceptual hierarchies common in legal systems are packed with value, and legal education is, for the most part, about learning how to identify and advocate priority among such values.

"Post-modern" theory (or anti-theory) has done well in emphasizing the complexity and depth through which value affects the scale, sequencing, and meaning of linguistic and legal events, but questions remain that are largely scientific. How does it work? Is there a pattern to this influence?

4. Our work on the stratification of moral and intellectual development may help us to understand language, but the key, it seems, is including "deity" as an imaginary placeholder in a model of consciousness.

It may prove useful, to begin with, to trace the path taken to this point—from a discussion of science, then the intellectual, moral and now the political. The structure of scientific theory (object) is a way of expressing or imposing value upon an exterior (other), i.e., to suppose it comprehensible requires the introduction of a value in its comprehension. In like manner, intellect is driven in the formation of object by interest, interest is chosen by identity, and identity responds to deity. As the arrayed ideas below suggest, value permeates being, and seems to have direction and continuity.

Value/Existence
Object/Other
Interest/Object
Identity/Interest
Deity/Identity

The continuity is important, because it implies sequential order to consciousness.

Deity/Identity/Interest/Object/Other

Moreover, this continuity suggests a sequence from Deity to Other like one from interior to exterior—Deity/Identity representing the interior mental experience, and Object/Other representing the exterior and remote mental experience. We have taken "interest" "( )" as a point of division between "identity" and "object" separating interior "→←" and exterior "←→." Visible here is a symmetry, of sorts, between other and deity, both representing the termination point of a continuum of experience.

Other←(→Deity←)→Other

By supposing other to exist with a structure similar to our experience of it, we are encouraged to clarify theoretic viewpoints—objects—thereby ready to record agreement or disagreement with them.

The practical need to maintain sensitivity to other is particularly acute when "other" references other human beings. The same could be said about worship. That is to say, by worshipping deity—a presumed basis of moral good, one acquires greater sensitivity to influences useful to self-cohesion. We have, in a manner of speaking, balanced the equation evident in a map of consciousness by inserting "deity" as an imaginary supposition, identifying an inner boundary to complement an outer boundary.

Using the sign "*" to indicate deity and "++" to indicate other, we have drawn a picture which may be of some use in describing language.

$$+\leftarrow(\rightarrow*\leftarrow)\rightarrow+$$

This represents an assembled portrait of consciousness, the validity of which depending entirely on its usefulness in explaining political discourse—i.e., the way human beings influence each other through language.

In a few steps, we have reached the point of argument for deity as an internal (subject) as opposed to external (object) presence. The diagram recollects an argument referred to as "the ontological proof of the existence of god" made famous by St. Augustine and Renes Descartes. The argument is—in essence—because we have an experience of perfection, perfection must exist. "God" is that existence. The argument has been criticized by a number of philosophers as an equivocation, i.e., that the existence as an idea is a different type of existence than existence as a conscious being.

Neither of these points of view is very satisfying. The criticism of Descartes' argument fails to address what "existence" is. Descartes' argument is influenced by the assumption that the existence of thought is proof of an existing subject that experiences the thought—an identity. Since ideas become better within a learning process, ideas emerge within an ethical universe in the process of discovery. The perfection toward which this emergence develops is God. God is, according to this argument, proven by the independent existence of an ethical universe.

The statement "I think therefore I am" has a strategic similarity to "I am better therefore God exists" because both call attention to the logical antecedence of a subject in all experience—a thinker of thoughts and an improver of improvements. It seems, therefore, that the "ontological" proof of God's existence is not refuted by God's non-existence as object because existence as object was not the intention of the argument.

Yet the "ontological proof of God" does appear to overstate necessity. The conclusion "must exist" is not the same thing as an experience of that existence. It is an assumption of constancy in value—of good—which supports developmental gradients. We do not experience that constant, but have a sense of the direction from which it comes, just as a wind vane's adjustments prove the wind's direction. As such it isn't proof of God, but of what we suppose God to be—of deity.

It is "deity" rather than God that exists as subjective antecedent to an ethical universe in the process of discovery, and "deity" may not be more than a psycho-social emergence within a neurological matrix. But then again, it could be more than that, though reaching social agreement on how to prove or disprove propositions of this nature can be troubling. A focus on "politics" does not require a resolution of the issue one way or another.

A perceived symmetry in consciousness—"deity" is to "identity" as "other" is to "object"—was likely active in Descartes' thinking, as he is also famous for his distinction between the mental experience of objects and an external world which "proves" its externality by impacting the senses whether or not we want it to do so. The western world has since named this dualism "Cartesian" after Descartes.

What is more provocative in the Cartesian system is an asymmetry consisting in the juxtaposition of deific existence—an internal source of influence—to objects outside of our perception of them—an external source of influence. They are asymmetical because they are not interchangeable, and establish a range between them in which gradients or levels of consciousness can be placed. In drawing this comparison, one engages in a geometry of consciousness that "locates" experience within a continuum. The more diversified and structured objects are called "external," which phase into unstructured valuative states called "internal."

The Cartesian distinction between an inner and outer division of reality is more apparent when other includes a group of other persons. The existence of or resistance to validation of our assumptions about reality is to a large extent tested through conversation with other persons, and it is difficult to imagine an outer reality without having been challenged by other persons with whom we collaborate in the management of a resistant material world.

This idea—descriptive of how other persons become involved in our orientation to an outer reality—is something with which we are highly and consciously aware when that reality becomes the reality of social collaboration and organization. The truly fascinating thing about another person is that the person is "other" to our experience, and thus known by inference, but engages one to imaginatively reconstruct their presence empathically, as if they weren't other. This is generally regarded as a social skill essential to group solidarity.

Therefore, an irony of politics is that what is "other" to one is—to the other—a subjective experience inclusive of object, interest, identity and deity. The diagram of an inter-subjective reality is more like this:

$$\text{Deity (1)} \qquad \text{Deity (2)}$$
$$[+\leftarrow(\rightarrow*\leftarrow)\rightarrow[+]\leftarrow(\rightarrow*\leftarrow)\rightarrow+]$$

Deity is an essential part of the "otherness" of other, which is to say, the deity of other is something which stands them up as an autonomous existence. With some time—and work—some consensus is apt to evolve between deity (1) and deity (2), which likely reduces fear and resistance in socially collaborative activity. At the point where the divergence is seemingly eliminated, there is apt to be just "deity." The question of political freedom is whether there is an ethical posture (value) that accommodates diversification of deity, or whether the elimination of the same is a necessary condition for peaceful social relations.

5. Before progressing much further, something must be said about the "point of view" communicated by this diagram. A blank page provides a nice medium for the assembly of ordered complexes because a page—or a line for that matter—is a continuum of places on which to locate events. But how is the mind like a blank page?

There is an attempt here to examine mind from the inside—which is the only mind we know of. The point of the diagram is to illustrate that freedom consists in large part in how the experiences of looking outward and inward equilibrate—particularly in competitive social environments. A human being has some choice over what that will be, but makes those choices while managing porous mental borders.

The arrows signifying inward and outward directions are states of attention.

$$+\leftarrow(\rightarrow*\leftarrow)\rightarrow+$$

It identifies five regions, but only three of these represent conscious experiences.

$$1\ 2\ 3\ 4\ 5$$
$$+\leftarrow(\rightarrow*\leftarrow)\rightarrow+$$
$$1\ 2\ 3$$

Discussion of mind in these terms is bolstered by evidence of qualitative discontinuity recorded in bureaucratic organization. Work organizations appear to be constructs which simulate a living organism, and are complex enough that one can devote a lifetime to growing into roles within them. Theories we have examined in the second and third chapters seem to validate a five part progression—five completing and restarting another.

Yet of the five places of mind we have mapped above, two are there because we suppose them to be there—a strategic orientation to certain unknowns with which we constantly interact. They can be projected into "roles," adapted to social organization, but with a living organism they are merely imagined. As noted in the first chapter theories of consciousness, both modern and ancient, propose tripartite models. This picture may therefore suggest common ground between three tiered theories of mind and language and five tiered cycles visible in human development.

It is hardly surprising that bureaucratic organization and other complex and politically charged human assemblies might break into five tiered systems composed of four different information processing modes. Our study of intellect strongly suggested that coming to know the world is a process of "identification" with and "distinction" from fields of information. Such objects may, in a manner of speaking, be regarded as replicants of consciousness because our coming to know them is similar to coming to know ourselves, that is, until we arrive at the understanding that the productive output of work evident the formation of organized social systems (among the many objects we create) resemble but are not the same as our individuality as conscious beings. That realization most often seems to come out of nowhere, as if God spoke, as a moment of distinction.

What we end up with is a model or map of mind in which a hierarchy of conscious experience ontologically defined as three places—identity, interest and object—effectively cleaves mind into internal and external sources of influence, each known only by their effects. The content of conscious experience changes as we change emphasis and "move," so to speak, between states of identity, interest and object. This "movement" includes our sense of existing on an asymmetrical

On Freedom: Political Freedom

continuum in which one type of experience can be distinguished and located. The process that governs this organization is what we think of as "language."

Human institutions of all kinds are evidence of an ordered continuum that is removed inwardly from a location we associate with a space/time continuum. Understanding those institutions is, in a number of important respects, a matter of locating their assembled constituents in relation to a continuum of places wired into the human psyche. Inasmuch as those places function as universals, they act as a dimensional axis from which locational coordinates are drawn. This axis would run perpendicular to the succession of events we refer to as "time"—immersion is represented by + and removal associated with *.[17]

One might imagine that ←→ and →← represent points of extension along the axis, separated and connected in thought by interest ( ). The sign →← points or indicates to * and the sign ←→ indicates to +. However, while this use of symbols and geometry makes a nice map of a continuum of mental spaces, it is not a satisfying description of the experience of continuity. It seems that the experience of continuity does not lie in the establishment of clear divisions, but oppositely, in the blurring of division.

Researchers interested in the physiology of the brain frequently remark about the intricacy of interstitial webbing between neurons, and systems of neurons, and of how that may allow us to perceive continuity in space and time. Take a line segment +* below, and ask what allows one to know that + and * occupy a common space.

Recollection and aggregation occur because there are intellectual processes that have enabled that to happen. It is significant that * does not just disappear when the eye travels to +, but that they are held together and integrated into one space. There are shifts of focus, of course, but such shifts are embedded within a manufactured continuum. Movement between mental spaces in language is not appreciably different. There are shifts of focus, but such shifts are embedded within a continuum of places.

Let us suppose, then, that there are three of five experiences that fall within conscious experience—Identity →←, Interest ( ), and Object ←→. Points of termination which are not directly experienced, but supposed are + and *. All five of these entities put together along a continuum signal order, or rather, of being closer to or farther away from points of termination.

$$+ \leftarrow \rightarrow (\ ) \rightarrow \leftarrow *$$

While we treat them as entities, their existence is a matter of relationship to the larger continuum of places which together form consciousness. The line is placed there to remind us of something—to stand for something. The line does not really exist, but is there to reinforce an order of placement. The same is accomplished with an overlapping matrix of experiences, of experiences which include their adjacent places. In a system having "continuity," each shift to a new place retains part of the complex that preceded it.

$$+ \leftarrow \rightarrow (\ )$$
$$\leftarrow \rightarrow (\ ) \rightarrow \leftarrow$$
$$(\ ) \rightarrow \leftarrow *$$

Returning then to the metaphor of interior/exterior, we are able to draw a figure that bears some resemblance to a living organism, reflecting vacillation between external and internal sources of reference. They do not—and cannot—exist as solitary entities, but are each complex mental phenomenon with a different point of focus, having points in common with what preceded transition. The pictures draw levels of language, and sequencing suggestive of a wave.

$$+ \leftarrow (\quad) \rightarrow + \quad \text{Semantic}$$
$$\leftarrow (\rightarrow \ \leftarrow) \rightarrow \quad \text{Syntactic}$$
$$(\rightarrow * \leftarrow) \quad \text{Presentive}$$
$$\leftarrow (\rightarrow \ \leftarrow) \rightarrow$$
$$+ \leftarrow (\quad) \rightarrow +$$

In an intellectual endeavor, we are engaged in the comprehension of objects, which may and often does include other persons. In a political endeavor such comprehension is only part of something larger. One comprehends other, but is also aware of being comprehended by other. One is influencing others to adjust, and being influenced to adjust to them. Almost by definition, this process describes a game, because it requires the adaptive participation of other persons.

A vacillating movement suggests activation from two sources of influence, one environmental (other) and the other existential (deity). Given that in a social environment "other" refers to the influence of other people, it would seem that we are gazing at a model of dialogue, of conversation. Other poses a challenge of sorts to mind, which evaluates the assertion and status of other, and existence reciprocates—casting influence outward to other.

Movement along a continuum is activated and rehearsed within a socially collaborative—and competitive—environment. It seems in one sense innate, as though we were pre-programmed to do it, but is activated socially, bearing attributes unique to a society which will be explained in more detail. There are at least three points which must be clearly expressed before the effect of a language model on political philosophy can be appreciated:

(1) One level of language shares mental space with other levels, and their differences are differences of attention.

(2) Movement between levels of language is visible in a shift of attention, and the release of engagement with a portion of what it shares with other spaces.

(3) The activation of deity in a game of language is the defining characteristic of political communication. This activation requires the assumption of a third tier of language denoted as Presentive.

6. Take the act of recognition of a human face, and upon a few moments contemplation it is easy to see that much goes into the statement "that's Greg." Greg's friend may have an entirely different reason for looking at Greg's face than an artist who has been commissioned to paint a portrait of Greg. Both are engaging the lines and colors of Greg's face, and their emotive relationship to those lines, but there is a world of difference in what they are doing.

Spoken language is often divided into a "semantic" and "syntactic" element. One addresses the meaning of words or signs and the other addresses the structure by which they are organized. At first impression this suggests a distinction between the referents of the words one uses, and their grammar. Typically when we use the term "organization," we are referring to a syntactical aspect to experience. However, now seeing syntax embedded in a model of language that includes presentive and semantic aspects, it is easier, perhaps, to see organization as an event within a process.

The overlapping nature of these events, while obfuscating distinction, suggest that the obfuscation of distinction has an important role to play in language. Semantics is more than that to which a given single word refers, because groups of words and groups of groups may have a meaning attached to them, all of which is the proper subject matter of semantics. Syntax may examine a single word, but the examination would likely reflect an interest in the way that word is put together so that it may stand out among others. The meaning of a word or group of words is a "semantic" interest, while the use of organization to facilitate meaning is a "syntactic" interest.

We may suppose that syntax is a higher order of analysis than semantics. This is due to the fact that syntactic rules typically apply with constancy to many semantic events. Some doubt was cast on this idea in the logic of development, in that the constancy of expectancy, a semantic interest, underlaid the methods for finding it. A given expectancy, however, lasted only throughout the duration of a cycle, and yielded to a new expectancy that would, over a period of time, become involved in the same developmental cycle deploying the same methods. Thus to say that this cycle was constant, while meaning was context bound, is to place the syntax represented by the form of that cycle on a different level.

We may ask whether there is, in fact, only one great expectancy, with many varying expressions of it. For this reason, it may be confusing to attempt to establish hierarchy in the distinction of meaning and grammar. It is more productive to focus on what shifts of emphasis occur as we attempt to communicate, i.e., whether there is a moment within a communicative act where one is more preoccupied with the organizational rules and parameters of the idea they are trying to express than with the meaning of the idea.

A "noun" is a group of sounds that stand for an object shared in thought by two or more persons. That object is a structured idea, having a set of characteristics that define it. These characteristics are what the object exhibits to our senses, and can be properly identified as adjectives, i.e., descriptive features that attach to the existential verb "is." We use "noun" and "verb" in much the same way we used "term" and "predicate" in logic. The only difference is in emphasis. In logic, emphasis rests upon the object. In language, emphasis rests upon a social process that results in shared experience.

A tree "is" a large rooted plant with branches. A "verb" is a characteristic of a particular object mentioned that places attention and significance upon it. The tree "is growing" is the same as the tree "grows." One could backtrack slightly to the expression "that is a tree," which is to say "that" (noun/term) "is exhibiting features common to trees" (verb/predicate). But the situation is the same.

The fact that an idea consists of a union between noun and verb (or term and predicate)—because it addresses the way ideas are structured—is a matter of syntax. That such an idea may or may not be true, because there is a reality beyond its formation that may or may not stand in agreement with the proposition, is a matter of semantics.

The phrase "that is a tree" or "the tree grows" has no semantic meaning without the identification of an interest which supports the effort going into the statement, that could include any one of the following:

> You need to learn English — "that is a tree."
> I need to verify my perceptions — "that is a tree."
> You need to understand what trees do — "the tree grows."
> I need to cultivate shelter — "the tree grows."
> We need to understand Language — "that is a tree" and "the tree grows."

The interest supporting the act of speech is what establishes the scale of the structured event, i.e., whether the communication is a single act, or part of a conversation or extended presentation. It also determines to whom the benefit of the act is intended. Sometimes one tells things to others in order to maintain a relationship that can vary in complexity, and sometimes one tells them things in order to delimit or destroy a relationship. For this reason, the evaluation of the semantics of a given act of communication can be a challenging mental exercise.

Though one speaking intends a meaning, and one listening intends to gather that meaning, it is subject to debate between the speaker and the listener. Who decides? The one speaking may think they knows what they intended to say, and perhaps as a courtesy the listener should defer what is later offered as a statement of intention. But the statement may in fact have been responsive to something the listener said before, made for the listener to hear, and the speaker may not know him or herself that well after all. Meaning, it would seem, reposes somewhere in the process of developing agreement between one and other.

A distinction or separation of a semantic from a syntactic level of language was represented pictorially as follows:

$$+\leftarrow(\quad)\rightarrow+ \quad \text{Semantic}$$
$$\leftarrow(\rightarrow \ \leftarrow)\rightarrow \quad \text{Syntactic}$$

The diagram argues, in essence, that semantic and syntactic attention shares mental space in the sense that both modes of attention engage "interest" and "object." Semantic attention works with intention (interest) and manipulates images (object) to convey meaning (other). Syntactic attention works with intention (interest) and the manipulation of images (object) with an emphasis on the pattern or way in which those images are organized (identity). The difference seems to involve a different range of mental spaces, even though the space content overlaps.

The semantic level includes a preoccupation with other, which the syntactic level does not. The semantic level "$+\leftarrow(\ )\rightarrow+$", in effect summarizes what we, in a discussion of the logic of work, identified as the components of intellect, i.e., interest, object, and an assignment of value to an event of discernment which affirms connection to or agreement with other. The notation summarizes the argument of intellectual freedom. Truth exists, as interest exists, but it exists by way of agreement between interest and object subject to test against other.

If Language is the medium by which identity forms interactively with other, and if language by definition includes a semantic event, then the status of this event establishes a foundation for the events that follow. That is to say, a theory of language builds upon the connection we make to others, and must, therefore contain within it the elements by which a connection to other is made. The acknowledgment that others exist and present an influence we manage, and to which we adjust, removes us from the imaginary and places us within a shared "reality."

Language engages issues of validation and consensus that certainly deepen our confidence in a reality that exists independently of what we want it to be, a reality that includes conscious agencies such as our own—with their own needs, interests, beliefs and deities.

7. We have, until now, said very little about movement within this continuum of mental spaces. It may have already appeared obvious from

a model of consciousness that vacillates between points of influence deemed "independent" of experience, and of arrows suggestive of where attention within a given mental space is cast. That was not to suggest that the space between the extremes of deity and other was passive or symmetrical, but quite the opposite. They serve as active and asymmetric processors of influence. Our individuality, our unique personal signature on life, is manifest in this process, an ordering process.

Along these same lines a more mechanical question lingers, i.e., how is movement within/among the spaces of consciousness experienced and recorded? Is there "direction" to this movement? The reason for raising the question here—at this point and in this way—is to show that the experience of movement is the experience of language.

One way to think of movement is as an accumulating history, so that movement is recorded in terms of growth. Language, however, is not as much about the recollection of experience—though remembering dialogue is a part of the experience—as it is about shifting states of attention. This phenomenon distinguishes the experience of language from the experience of growth, i.e., that we shift our attentions without attempting to coordinate or combine them together.

Consider the transitions below, one recording the expansion of space, and the other recording transition from one space to another:

$$(\cdot)\cdot \text{ to } (\cdot\cdot) \text{ versus } (\cdot)\cdot \text{ to } \cdot(\cdot)$$
$$(\cdots)\cdot\cdot \text{ to } (\cdots\cdot)\cdot \text{ to } (\cdots\cdot\cdot) \text{ versus } (\cdots)\cdot\cdot \text{ to } \cdot(\cdots)\cdot \text{ to } \cdot\cdot(\cdots)$$

One might suppose that in the first situation that there is little, if any, sense of mobility between spaces, as it appears to be one space getting larger. In the second case we have added a mobile and adaptive feature to movement because one has—in at least a sense—released attention from the first space and moved from that to another. Being able to release one's attention to a given mental space allows one to process diverse sorts of information without becoming smarter.

In speaking, there is movement from syntactic to a semantic level of attention. There is quite a difference between the preparation of an utterance where one is focused on structure, and the use of structure after one has shifted their attention on the meaning toward which resultant structures flow. In the syntactic moment one is engaged in strategic deliberation and assembly that fits and resonates with their personal aesthetic. But self-conscious reflection over one's own fit in the world is released in the semantic moment, as one becomes focused on what they want and how to bring that to other—to make meaning.

Hearing an utterance reverses the order of events. One takes in the meaning of what is coming from other—structured intention of an autonomous being. But this does not terminate what it is to listen, as one's attention shifts from the otherness of the communication, and undertakes evaluation and measurement of it—i.e., what it states about the identity of the speaker, and how that compares with his or her own identity. That evaluation cannot occur in a vacuum, but involves the relation of structure, interest and identity compared to a sense of beauty—narrative. Is the other's narrative in accord with the listener's? If it hasn't registered, then it is fair to say either that communication failed, or that one hasn't listened to the other.

Knowing how to release from preparatory considerations and enter the flow of execution is no less important than any other aspect of language, and if it did not happen, language would not exist. The same can be said of listening, i.e., of ineptitude in one who fails to step back and accurately assess the personal form of other being expressed by comparing with their own.

Looking at a picture of this process, it is apparent that the key changes in the movement to and from semantic and syntactic levels of language are the engagement and/or disengagement of other and identity. The constants in this movement are object ←→ and interest ( ). Direction is a matter of whether one is speaking or listening.

$$+\leftarrow(\quad)\rightarrow+ \text{ to and from } \leftarrow(\rightarrow\leftarrow)\rightarrow$$

In trivial or casual conversation, the shift occurs so rapidly that the discernment of shifts of this sort is purely intuitive. Fluency in a spoken language, by definition, represents sufficient familiarity with the words and grammar of a language that there is no hesitation between thinking and doing. Yet as one fluent in a language fails to communicate, they are apt to proceed more deliberately, to the point of speechlessness. This suggests a strategic element to language in very simple discourse that is less obvious, not less present, than in larger or more formal political exchanges, or in one who is attempting to learn a new language.

A significant aspect of the arts of speaking and listening is in fitting communication to the opportunities and restrictions of the medium in which it occurs. An over-managed project makes it difficult to adapt to unforeseen or idiosyncratic events, such as a resistant audience. A courtroom advocate knows that preparation does not consist in the memorization of a script, but in an orientation that allows one to respond convincingly to challenges. A novelist knows that one of the keys to writing is the creation of a situation where the novel lives in his or her mind, and speaks to them as they write it.

As the argument for intellectual freedom is depicted by semantic event, the argument for moral freedom is summarized by the depiction of a syntactic event $\leftarrow(\rightarrow\leftarrow)\rightarrow$. In the fictional narrative, one's vigilance over the truth of factual statements is suspended, as the author and reader probe a story for its moral statement. Facts yield to that statement because facts at that point are not there to be factual, but to reinforce a setting where interest and identity agree. Moral judgment thus marks a shift from intellectual judgment—releasing other in order to engage identity.

We may likewise view a shift from moral to intellectual judgment as a shift toward a semantic level of language. The moral engagement of self-consciousness preoccupied with social and practical boundaries seeks orientational comfort—a sense that who one is can exist comfortably with the practical limitations of the world.

This is a radical position, i.e., that every communicative act—no matter how trivial—shifts to and from an evaluation of one's fit into a social environment. But it comes with the association of language with politics.

It is, perhaps, the hardest part in understanding what language is, because a model of language in which human beings blend in and out of introspective and assertive frames of mind is messy—but no less than the purposes being served by language. There is much more at stake than the exchange of ideas. The maintenance of relationships is no less important, and we have evolved a mind capable of combining these functions in powerful ways, while moving among them.

8. Which brings us to a third level of language, labeled above as a "presentive" level—so called because one's attention there moves into being in relation to others, "presence." This event in discourse releases structured objects and concepts and dwells in a sanctuary of sorts where deity is permitted to interrupt the momentum of assertions imposed by other.[18]

Relatively innocent communications convey information about the relationship of one person to another. Even when the intent of the communication is to validate the identity of the other, there is a question of whether that perception is accurate, or in keeping with how the recipient knows his or her self. To say to someone "you are wonderful" can raise issues. Do I need to hear it, are there doubts about me, or is it your role to evaluate me?

The moment in which one assesses how such assertions affect or sit with their moral identity fills discourse with a dynamic inter-reflection of value. We can—with some effort—construct symbolic dialogues where such ethical exchanges are muted, or eliminated. It is difficult to imagine that such a dialogue would persist for long, given the human interest in giving and receiving ethical validation to others. It would be a wholly artificial language designed to circumvent an essential substantive ingredient of human interaction. Even there, it only mutes this aspect of language, as it might appear—after substantial consideration—that language cannot exist without this functional aspect.

An author of science fiction once satirically described a planet of simple beings, which, like a form of plasma, lined the caves of Mercury, and took turns giving and receiving sounds, which by translation conveyed "I am here" and "so glad you are." Likewise in human communication, underlying the intricacy of the messages sent and received, one seeks to challenge and validate others, and receive the same from them.

To further emphasize the satire of the story, the author called this species on Mercury "Harmoniums" because they exist—in contrast to human beings frequently at war over limited resources—only to validate each other. Political discourse within our species appears to involve much more in the form of "choice" over what will be deemed acceptable in the various assertions that others make. An ethically unstructured and permissive orientation is sometimes referred to as "liberal" while a structured and authoritarian orientation may be referred to as "conservative."

In either case—liberal or conservative—there is a point of engagement with deity through which ethical leanings emerge. It is useful to have some sense of the mechanics of that engagement if for no other reason than to anticipate how collaboration and conflict might flow from such engagement.

Suppose, then, that a presentive event exists that disengaged from a structured object, and that what remains of conscious experience is interest, identity, and an active receptivity to deity—an event depicted as (→*←), where one consciously situates in order to receive influence from conscience. In ordinary dialogue, this event could be momentary, only enough to activate a response from deity, and may even appear as a reflex. But in other contexts the challenge posed may be stronger, and activate a prolonged and contemplative reference to deity. Prayer and meditation—arguably a contemplative withdrawal from a structured engagement with other—are strategies involving conscious effort to prolong or intensify one's receptivity in a presentive event.

The meditative contemplation of deity and its relationship to self-interest, is a private experience excluding other and object, but prompted by the influence of other. As one arrives at the sense that their interest conflicts with another's, the "who" of what one is →← is in doubt. That is to say, the presence of conflict evokes concern over the validity of this identity, and whether one should acquiesce to the conflict by adjusting one's identity, or resist the other who is the cause of this conflict. That issue is resolved through authoritative reference to the deity that reveals the "what" of what one is. The process then returns to a syntactic level, where strategic considerations in the affirmation of identity—in context—are arranged.

Now we can get much closer to describing what happens when there is disagreement over what is right—the situation described earlier as political conflict. This is quite different than what we have earlier referred to as disagreement over whether to do what is right—which describes moral conflict. Let us say we have a situation where there appears to be a divergence of conscience between two participants in a conversation—two different manifestations of deity.

$$\text{Deity (1)} \quad \text{Deity (2)}$$
$$(\rightarrow * \leftarrow) \quad (\rightarrow * \leftarrow)$$

There appears to be a limited set of options between them.

- One can persuade the other,
- The two can coexist with an agreement to avoid alignment.
- The two can alter their sense of deity to eliminate the conflict, or
- One can eliminate the other.

The first three options suggest a process of spirited—though not necessarily combative—political negotiation unfolding as a language game. It is not difficult to see any number of combinations of the first three options forming, and mutating as political dialogue unfolds. What these options have in common is recognition of the privacy of one's relationship with deity, and a willingness to avoid reference to deity in terms of absolutes. In a world of absolute deities, one is limited to social association with persons whose deity is very similar to their own, and is otherwise bound to the fourth option—elimination.

The assertion of an object deity, that is, the extrapolation of "God" out of one's existing sense of moral good is prone to intolerant attitudes because they are, by definition, representative of a fixed moral position incapable of negotiation. Thus, the event depicted by (→*←) is descriptive of political freedom because it acknowledges that our relationship with deity is a process of adjustment.

It is quite possible—and in most modern societies quite acceptable—to believe in an absolute ethical presence or "God" and regard one's own manifestation or expression of that presence, "deity," as an evolving interpretation of it. This occurs through the acknowledgment that deity is obfuscated by limitation present in one's sense of self and his or her range of interests. Thus if God commands "you exist as my perfect idea," it may be heard as "you exist to serve" or "you exist to dominate," depending on what sort of influence identity is prepared to receive.

The reception of a message would likely affect whether one acquiesces to or resists challenges to interest. To hear this command as "you exist as animal" or "you exist to be free" also has implications with regard to how consciousness will endeavor to reseat its identity when provoked by other.

When another presents a self-image in conflict with one's own, such as to say "I deserve the prize and you don't," this message impacts at a point, and an idea about who one is emerges. It may be, "you are better than I" or "sometimes you are better" or "you were lucky" or even "you cheated." Underlying each of these responses is a message about what one is, which could be "we are not created equal" or "we are equal but must take turns at winning" or "we are the fools of fortune" or "we are the victims of mischief."

9. There is some uncertainty over the use of metaphors. In this case we use the term "language" to describe "politics" while occasionally suggesting it is the other way around, and that "politics" describes language—a game of influence.

While language seems to have political features, much of what we typically associate with politics is not what one identifies with spoken communication, but with the management of relationships in government. In order for language to be an appropriate descriptive system for what one would call "political freedom" it would help to show how law is communicative, and that its communication occurs with the structure identified with language—as a cycle of semantic, syntactic and presentive events.

Let us say, therefore, that language, by definition, involves a communicative relationship between any two or more persons. Can we also say that politics is like language because government is a "person" and that each of the subjects of a government is in a communicative relationship with it? This question, it seems, raises the issue of whether an organized collection of persons can be regarded as a person, as a morally autonomous individual speaking as if to offer its existence for evaluation, subject to a response from its citizenry.

Perhaps it isn't necessary to resolve that issue, because government speaks through individuals. Suppose instead that the mind of a government is in the mind of one appointed by it to speak, and to speak on subjects germane to the relationship between government and citizen—of laws and obedience. In that sense, politics typically associated with those bound in a governmental relationship is better regarded as a variation of a language game. The limitation of the subject matter—along with the authoritative status of its participants—shapes it into a variant of language.

The subject of communication between government and citizen—or sovereign and subject—is obligatory conduct. Government asserts and reinforces obligation by reposing authority in some over others. The others may either accept or resist this obligation, depending on the persuasiveness of that authority. This relationship, some assert, is all that is needed to establish law. Theorists who argue that this is so subscribe to a philosophy called "legal positivism."

Another school of theorists who dispute this philosophy find it remarkable that those who assert such rules care about whether the rules are moral—as do those who choose to obey them. They advocate what they call the "natural law."

Positivists meet this objection not by denying the efficacy of moral sentiment, but by insisting that moral sentiments do not change what law is. They are merely descriptive of one among many persuasive sanctions that produce obedience. While law may be less effective when it conflicts with moral rules, it can theoretically exist without them. This is offered as proof of legal positivism, i.e., that coercive rule of sovereign over subject is possible.

The positivist view further responds that natural law is unnecessary to resolve conflict among laws, because such conflict is resolved by a set of rules through which that selection is made. These belong to a super-ordinate set of rules often referred to as "rules of procedure." The positivist argument avoids the objection of the proponents of natural law by pointing out that while moral rules may help to resolve conflict among laws, they are not necessary.

The positivist school is often wrongly confused with an authoritarian political perspective because it emphasizes the fact that law may, theoretically, work with or without moral sentiment. However, positivism is more ethically nuanced than it may appear. It arose in a period of history when diverse and competing moral systems were being assembled into nations and empires, and the assertion of a morally neutral order was preferred to religiously fueled conflict over which moral system might rule.

The idea that we might form a social contract, of sorts, agreeing in advance on how to resolve disputes over obligation seemed preferable to the chaos associated with the establishment of preference between deities. Positivism was, in many ways, an expression of respect toward the motivational power of moral sentiments spurred by variance in human conscience.

At issue, therefore, between legal positivists and the promoters of natural law, is whether rules are enough to keep the peace. This issue is, by translation, similar to issues of completeness encountered in logical systems—i.e., whether a system of axioms is a sufficient and satisfying explanation for a rational system. Here we encounter a similar problem of regress examined in both logical and moral systems, for clearly: (1) value is essential to judgment and (2) objects (rules) do not confer value on objects.

Perhaps it is less clear that a two tiered political system broken into "substantive" rules (rules of obligation) and procedural rules (rules which resolve dispute) has the requisite completeness to constitute a real system of laws. The placement of three tiers of language beside a two tiered system of law may clarify the problem to some degree.

<u>Language</u> <u>Law</u>
Semantic Substantive
   (Meaning)   (Obligation)
Syntactic Procedural
   (Organization)   (Conflict resolution)
Presentive [Constitutive]
   (Identity)   (Authority)

Given functional similarity between semantic-syntactic and substantive-procedural levels of language and law, and the observation that law is a form of communication, we are drawn toward an apparent void where a third level of law is unstated. Is that void merely a rhetorical device, or is it a real argument supporting the assertion and clarification of a third level of law? A third level of law which enacts systemic goodness, it seems, resolves the long standing debate noted above between natural and positive law.

While there is strength in the proposition that law can and often does function without moral support, positivism must admit that the resolution of controversy over what the law is essentially requires an evaluation of what the law ought to be. While the moral significance of law may not be easily seen, it is nonetheless revealed in cases of controversy.[19]

That argument is similar, in fact, to the one we visited earlier in support of a presentive event in language, i.e., a conflicting ethical presence asserted by other provokes a responsive act of clarification. Likewise, a sovereign assertion of obligation that produces a sense of conflict in a subject is apt to be met with resistance. Such resistance—whether covert or overt—is a communicative act that forces adaptive responses in game behavior. Human resistance to an ethical assertion is, by definition, a "controversy."

Much of the confusion between the schools of positive and natural law derives from disparate views on the timing and frequency of ethical intervention in the administration of justice. There are good reasons to promote stability in the application of rules of obligation, and good reasons to resolve disputes fairly. But the "goodness" of these reasons is an ethical assertion, which means there may be reasons to amend substantive and procedural rules.

One might therefore say that the positivist school "loses" the debate, given that any reasons for the implementation of a rule based system extend from an ethical platform. But that would be premature. The positivist may well be satisfied with the assertion that there are rule based aspects to law, and that the importance of positivism was never to exclude the whole universe of ethical influences on the administration of law, but to acknowledge and define a rule based dimension to politics—and thereafter refer to it as "law."

This would be an attractive refuge for positivism because it imagines a universe of rules in which we can maintain distinction between law and politics generally. The problem with that view is in the simple fact that legal judgments are—as all acts of language—compound assertions of fact and value. One may, from time to time, shift emphasis and focus, but the logical extrication of an object (rule) based world from a value oriented universe misses the epistemic reality of both logic and language.

A benefit of the positivist school is in its implicit mandate to pay attention to the structural aspects of law, so that ethical disputes may be effectively framed for resolution. Without a well formed syntactic event—an effective procedure—it is quite difficult to reach the point of real ethical controversy. This may, in effect, summarize the positivist argument, i.e., that indiscriminate intervention of moral sentiment within a lexicon effectively disables the proper use of moral sentiment to resolve controversy at the level where controversy genuinely exists. That is, in effect, what it means to establish what we refer to as "authority" in law suitable for the resolution of ethical controversy, i.e., to establish ruling principles for a system of laws, and an administrative level at which those principles can be finally implemented.

This principle applies not only to the formation of effective legal systems, but to any kind of organization we can imagine, so much so that persons who are not intuitively aware of and competent in this practice are not, as organizational participants, taken seriously.

The strength of a language based metaphor for the description of law is that it accommodates shifts of emphasis in the experience of law without placing a system of laws at risk of chaos. We can view law—the productive output of political behavior—as a dynamic process that promotes and responds to ethical postures through structure. In other words, the positive aspect of law is indispensable to, but not a complete predicator of, what law is.

10. Politics cannot be free—be what it really is—unless it effectively includes a presentive event. Such an event can be chosen, but only if there is an understanding of what it is, and what the foundational prerequisites for it are. Accordingly, there are a number of things that politically free societies do to effectuate political freedom. That they do these things can be regarded as fair evidence of the fact that political freedom exists, i.e., is alive in the minds of its participants.

There are things we equate with political freedom that may, in fact, be quite dangerous and disruptive when implemented blindly. Notable among these is what is referred to as the "vote."

A persuasive reason to resolve controversy through ballot and proxy is that it is a less expensive way of achieving orderly and peaceful relations than other alternatives which involve coercion. This is, as mentioned above, the ethical foundation of positivist jurisprudence, i.e., the avoidance of violence in the resolution of moral disputes. Yet peaceful transitions of authority are not the only thing of political interest that voting produces.

Voting changes the political dialogue from an active-passive exchange between sovereign and subject to an active-active exchange. Authoritarian political systems have quite a lot of dialogue with their subjects, but the value of the political system is measured by the receptivity and obedience of the persons under rule. Voting brings politics into closer alignment with language by creating a situation where authoritative figures are required to respond to assertions of other. Mutual and relatively equal influence between those who make and obey law challenges and activates conscience, and is thus apt to support more ethically complex and inclusive social institutions.

Political freedom therefore favors dialogue between government and its citizens—enabled by democratic voting processes. An obvious corollary of the vote is political activity designed to assemble others to vote similarly. This typically requires the sanctification of self-expressive media, as in spoken communication, art, music, etc. and the social assemblies such as political parties and action groups. Such media and those who assemble around them are apt to represent a very interesting mixture of self-interested advocacy, and advocacy designed to promote higher ethical principles. That is the nature of autonomous moral beings—i.e., constant activity in blending and re-blending personal and social interest.

There is quite a difference, however, between the communicative influence (speaking, organizing and voting) of a morally autonomous being who includes conscience in their expressive activity, and one who is obligated to refrain from imposing personal conscience in their acts. This is usually the case where one is being paid or employed to represent the economic interests of another. There is nothing unethical or immoral about assuming such an obligation, and to so act even where it may appear to violate conscience. The problem consists in the fact that one speaking under an obligation to refrain from personal reference to conscience is, in effect, suppressing exactly what is required to speak freely and to participate in political freedom.

While lawmakers may, as a matter of policy, couple the speech and organizing activity of corporate entities with that of autonomous moral beings, the rationale for doing so is perplexing—except perhaps because it is difficult to administer such a distinction, or because the wealth of private interests allow them in some other way to procure sanctuary for their political activity.

Possibly the most common criticism of democratically ruled governments is that they are corruptible. Since communication to mass audiences is expensive, political speech slants toward economic interests which rent access to communication media. These interests speak with a louder voice, and are—more often than not—incapable of including conscience in their communicative activity due to corporate obligation. They speak as if they were motivated by conscience, but the speech is not honest.

It is not, in fact, unusual for organized economic interests to appropriate democratic governments in their entirety. The strategy in doing so is in disabling free political discourse. This can be done without coercion, i.e., by evoking panic and terror, compromising and confusing the common meaning of language, the dissemination of false beliefs and opinions, and the stirring of resentment and anger between diverse moral perspectives.

Politically competitive practices that successfully disable the expression of conscience can then turn lethal. As in all games, as participants learn that benefits accrue to selfish and non-collaborative behaviors, the power of the vote to encourage and uplift political discourse is compromised by manipulation and violence. An aggressive minority would use coercive tactics to frighten a submissive majority. An oppressed majority might, upon assuming political control, make dreadful retributions.

It seems that the vote may—under a number of circumstances—fail to promote a higher ethical position in the expression of law. This is due to the conflicts apt to emerge from vigorous self-assertion, and a failure to anticipate and manage those conflicts. What we would expect from a politically free system is to enable autonomous moral expression, and to protect political discourse from degradation. There would be laws designed to assure fairness of access to the lawmaking functions of government. This might, with some irony, involve regulation of the apparatus of political communication, to introduce constraints in order to enable and amplify morally autonomous communication.

11. The vote may engage sovereign authority to listen and adapt to an electorate, but sorting through the noise of such dialogue, so that matters of importance will be effectively discussed, is another matter. Given the chaotic tendencies of unrestricted dialogue—especially in the conduct of elections—political freedom demands orderliness in the

formation and execution of rules arising from that dialogue. This is often referred to as "the rule of law."

The rule of law emphasizes coherency in a system of laws, from their inception to their application. That coherency is usually institutionalized in the form of a system of justice which interfaces with the public—bound by this system of laws—and served by a profession of persons specially tutored in the law. Their vigilance in the maintenance of a set of laws within a system of justice establishes a syntax and vocabulary which the public can address as if it were the product of but one mind.

It is less important that the law actually cohere than it is that those benefiting from the protection of law support and protect coherency—i.e., hold it as a value. Imagine speaking to an individual on matters of principle, but a supposition underlying the conversation is that no one is obligated to comply with rules arising out of the dialogue. The adaptive expression—the game—of such a dialogue is apt to be superficial, and could be limited to the repetition of platitudes. A system of justice elevates law to seriousness, which includes being effective and consistent.

A lexicon—a rule oriented hierarchy that sorts through gradients of importance—and an effective way of bringing disputes together, is necessary for legal discourse to separate from and focus upon genuine controversies. Without it, all gets lost in the details. In a politically free society, this hierarchy is typically present in the discourse of every government agency that legislates, executes and adjudicates matters involving law.

Hierarchic functions that serve this purpose are most noticeable in judicial institutions—"courts"—because these are tasked to resolve disputes. Not surprisingly, there are typically three tiers of judicial function—three types of courts—which comprise a system of justice. One of these makes adjudications of disputed facts, a second tier resolves disputes over the law that governs the fact finding process, and a third tier resolves authoritative conflict on who decides what the law is. Many jurisdictions refer to these as the "trial" court, the "appellate court" and the "supreme court."

In the minds of persons who work in court and communicate with each other on matters of evidence and law—usually attorneys—there are semantic, syntactic and presentive aspects to their communication. These, as we observed, are in legal parlance referred to as substantive, procedural and constitutive law. Significant to a system of justice is that its tiered courts each—the trial, appellate and Supreme Court—assumes authoritative roles comparable to semantic, syntactic and presentive events. A system of justice is meant to replicate or "embody" the mind of sovereignty.

Let us look a little more closely at a "trial"—a semantic event. A trial addresses interest in whether a specific person has violated an obligation—an application of substantive law. Though the emphasis of the event is substantive, it uses pre-selected rules of construction that manage the process by which that proposition will be assigned a conclusion, true "+" or not. A "trial" is a semantic event because its purpose is to assign meaning to a given proposition—which is here a proposition involving the status of an individual under substantive law.

The maintenance of "fairness" in a trial derives in part from a line of separation between those who argue a proposition (Advocate) and those who determine whether the proposition is true (Jury). The proceeding is gathered under concerns raised over whether a person has complied with obligation (interest), and is usually supervised by an official (Trial Judge) who determines the standard of obligation, and directs the investigatory procedure.

Semantic	Substantive
Interest	Trial Judge
Object	Advocate
Other	Jury

The legal equivalent of a "syntactic" event is what is often referred to as an "appeal." "Appeal" releases—in varying degrees depending on jurisdiction—engagement with fact, but newly engages dispute over whether the fact-finding event was properly supervised. This will include review of whether procedural requirements were met, which is apt to include whether the parties to the dispute were properly allowed to present evidence and argument, and whether the fact finder labored within a correct statement of the rule of obligation.

A review of this nature, of course, would not have been possible at trial, because the conduct of the trial is—in the judge's mind—self reviewed and corrected. But the trial judge functions within a range of what is acceptable and what is not—a field of tolerance governed by a sense of the identity of the law. While he or she might suppose that they fit within that field, review is delegated to one or more persons at a higher level, who review a transcript (record) of the trial. A shift occurs.

	Syntactic	Substantive	Procedural
Semantic	Identity	Substantive	Appeal Court
Interest	Interest	Trial Court	Advocate
Object	Object	Advocate	Record
Other		Jury	

Underlying the process of review is an awareness of the limitations under which the legal system must operate. There is no special science to finding fault in the work of others. Depending upon the scale at which scrutiny is directed, all human activity or no human activity is error. A syntactic event is aimed at fitting interest to a context involving certain limits in the time, priorities and media in which communication occurs. An appellate court likewise seeks to match the evaluation of interest with the functional parameters of a system of justice—the identity of the system.

Without question, a judge who sits on a court of appeal is influenced by conscience—as all persons are. He or she may well consider authoritative conflicts. Is the legislature or the executive in conflict with rules that precede or supersede legislative enactments? But on such matters, controversy is apt to be so deeply felt and expressed that a supreme authority—the conscience of the legal system—must be appointed. This represents not only a practical need to resolve controversy, but a need to enable the expression of conscience by allocating administrative authority for doing so.

While all litigants disappointed with their result at an appellate level have a right to ask a supreme court to look at their case, that court typically reviews only those cases where the controversy is worthy of such review. Supreme Court review arises where the branches of government appear to be in conflict, or where appellate courts render conflicting decisions. It does not attach as a supervisory right to individual litigants. A supreme court releases such obligation so that it may become focused on the clarification of the ethical basis of a system of laws.

	Presentive		Constitutive
Syntactic	Deity	Procedural	Supreme Court
Identity	Identity	Appeal Court	Advocate
Interest	Interest	Advocate	Authority
Object		Record	

We could partially suppress the intervention of conscience in the resolution of controversy by eliminating a supreme court or removing authority from that court to apply conscience to the resolution of controversy. There are many advocates, for example, of a policy of strict construction in the interpretation of law—even those laws which were intended to precede and establish foundation for the acts and decrees of legislative and executive branches of government.

The difficulty with an attempt to restrict such discretion is that it leaves a void in the law for which there appear to be no acceptable surrogates. When laws conflict, what is the basis for the resolution of the conflict, except to suppose that the law is intended as an ethical solution for complex social relations? The rule of law does not allow a judge to replace duly enacted laws with a personal morality, but it does support the resolution of conflict as if laws served justice as principle.

One solution to this dilemma—i.e., activating and enabling conscience while restricting licentious intervention—is to articulate principles in a document that serves as a highest law. A constitution that establishes a structure of government and the powers of its diverse agencies, together with some description of the objectives being served by that government, helps to empower a supreme court to interpret a system of law in accord with ethical principles embodied in the law. This enables the intervention of conscience while limiting what it can do.

There is therefore no such thing as "strict construction" when it comes to the task of defining and limiting the authoritative power of government because the rule of law is not about the elimination of value from politics, but the effective management of value in a system where many interests conflict.

Even a decision not to intervene is a form of intervention because it cedes authority to lesser courts—thereby multiplying the authority under which the system of justice functions and weakening the rule of law. This is, in practice, the fallacy of strict construction, in that it seems to prohibit judicial intervention where there is an ethical controversy in the interpretation of policies. Such prohibition does not preserve the rule of law, but to the contrary, corrupts it. The consequence is politically disabled sovereignty—a speechless void apt to be filled by more politically coercive, restrictive and closed systems.

12. Political freedom arises from an attitude that disengages one from a given moral system, and that values the diversification of moral viewpoints. Despite the most ingeniously devised systems for the administration of law, political freedom is a state of thought, and thus

requires thoughtful participants. While the politics of consciously selecting an aristocracy is apt to be divisive, the promotion and education of a sophisticated and inspired view of freedom might help to enlist capable individuals to leadership.

A society of intellectually and morally free individuals, while ready to achieve political freedom, may yet fail to do so because they are held by a sense of deity which excludes rather than embraces other identities. This normally occurs when this sense of deity, i.e., what one is, never really separates from who they are. What one is a mystery, but is revealed incrementally within an environment which allows adjustments.

We have examined two features of political freedom: (1) "the vote"—a mechanism which promotes dialogue between a sovereign and its citizenry, and (2) "the rule of law"—structural protections within institutions which assure the selective intervention of conscience in the administration of justice. We have noted that it is useful to codify principles and procedures that clarify the kind of interventions that are welcome. The content of such principles are usually called "rights" because they establish boundaries on the authority of sovereignty. There appear to be at least two theories of government that explain the existence of "right."

One of these emphasizes the need to arrange for the peaceful transformation of government. This view suggests that the only way to encourage consent to the dominance of a majority based legislature is to establish inviolable rights, i.e., that the protection of certain interests are necessary to promote enthusiastic participation in a process the effectiveness of which depends on universal inclusion. Politics is a consensus generating activity which empowers government. This theory is, more or less, an elaboration on the value we have already associated with "the vote."

The other emphasizes the identification of rights whose protection is fundamental to the enjoyment of life. This view suggests that sovereignty, regardless of how it is constituted, is attached to a goal of domination. The respect of fundamental rights is necessary to curtail the natural aggressiveness of government. The power manifest in a vote is only one among a number of such protections. Politics is a struggle of self-preservation against the influence of government. This theory is an extension of the value associated with "the rule of law"—i.e., the way rules instruct and constrain the discretionary assertions of government.

These views tend to support similar government structures, i.e., a democratic elections process and a list of protections aimed at preserving personal autonomy. The two theories tend to differ in their application to situations. The first is willing to qualify protections in a manner suitable to the promotion of consensus, and so welcomes governmental interventions which promote that objective. The second is generally opposed to entrusting government with the task of determining which qualifications are and are not suitable.

They are not the only theories of government circulating in the academic press, but they serve to illustrate how conscience can become important in making decisions about how to design and maintain a government. Much can depend on one's sense of what the human organism can tolerate and what it can't, and that depends on one's sense of what a human being really is. Divergent deities speak differently about rules that all regard as inviolable. Take the following familiar principles written into a codicil following formation of a federal system of government in North America in the late eighteenth century.

> The government shall not "establish" a religion, or prohibit the "free exercise" of religion.

> The government shall not prevent the "freedom" of speech, and of "peaceful" assembly.

> The government shall not engage in "unreasonable" searches or seizure of one's property or person.

> The government shall not take one's life, liberty or property without "due process" of law.

Certain phrases are placed in quotes because they literally quote from that codicil. They are also quoted because they are vague and remarkably brief, and seem therefore to acknowledge a significant void which would be filled over time. Consider the terms "establish," "free exercise," "freedom," "peaceful," "unreasonable" and "due process." These terms are as controversial in their meaning as are divergent theories on what makes a political system good or ethical. They are ethical principles designed to enable the intervention of conscience within an authoritative order.

If the theory emphasizes the value of representative government (first theory above) then one might, for example, view the first and second statements of principle as being similar or even identical—as the social organization and promotion of moral systems is essential to representative government. Under this theory, the third and fourth principles are designed to protect political minorities from being intimidated or marginalized by a victorious majority—i.e., not so much to protect them from harm but to insure political continuity of minority interests.

If the theory emphasizes the value of individual avoidance of government or individual liberty (second theory above) then the third and fourth protections would seem to be similar, or identical, i.e., to stop government—often driven to madness by immoral majorities—from harming individuals. Freedoms of religion and speech are merely an acknowledgment that human social relationships—inclusive of religious gatherings and political assemblies—are cherished interests worthy of protection.

In devising a theory about a set of rights that have significant importance to modern civilization, we are, in effect, developing explanations for what is good about them—about why they represent an ethical improvement on the past. In doing so we acquiesce to the argument that these rights are defining statements about what we are—i.e., morally autonomous political beings with shared ownership of government.

The theoretical examination of law as a species of language eases the sense of division among diverse political theories. Law engages individuals in a dialogue which challenges fixed "object" deities and forces tolerance and accommodation of a more diversified moral environment. The continuity of government and the protection of individuals are both enhanced by a broader field of play. The ideal of political freedom emphasizes enriched individuality through the preservation of other. This is more visible to those who appreciate similarity between law and language.

The "rights" enumerated above—religion, expression, privacy and security—are protected because political freedom requires the growth and participation of autonomous moral beings. This would prohibit government from interposing itself between identity and deity—an exclusion between church and state. Because what is being protected in this separation is a formative process through which deity influences identity, government restraint is not limited to religious issues. Political freedom requires that government remove itself from a position where it may inhibit the expression of or experimentation with identity.

Rights established under a constitutional government followed after centuries of cultural evolution which included religious reform, and scientific advancement. Some believed that a government could be scientifically designed to protect humanity against its weaknesses. The conception of "right" or "civil right" as a constitutional guarantee functions as an imperative that political systems operate within ethical boundaries that protect individual sovereignty, while allowing clarification of what those boundaries are. A concept of "right" thus stands out with (1) the vote and (2) the rule of law as a third distinct aspect of political behavior necessary for the establishment of political freedom. It is a way of formalizing and effectuating a presentive aspect of law (as a species of language) in very large and unmanageable political environments.

The world has progressed somewhat since representative democracies supportive of free markets were first conceived and implemented. There is some indication that ancient and modern civilizations are beginning to integrate—a process which is apparently distressing to political regimes which do not agree that "freedom" is an appropriate moniker for the transformations associated with this integration.

To date there is no effective technology for the conversion of ancient to modern political structures, and that too is a problem given that there is growing competition over natural resources in a world market. Though one might therefore say that it is as easy as teaching a child to talk, the statement is problematic for a number of reasons—a civilization is made of valuable inheritances protected and advanced by adults, and it was not that easy to learn to talk in the first place.

5.

Spiritual Freedom

I say to you, except a man is born of water and of the Spirit, he cannot enter into the kingdom of God. That which is born of the flesh is flesh; and that which is born of the Spirit is spirit. Marvel not that I said to you, you must be born again. The wind blows where it lists, and you hear the sound of it, but cannot tell from where it came and to where is goes: so is everyone that is born of the Spirit.

Jesus.

1. A theory is like a wave coming to shore. From time to time thought, motivated by what seem to be facts, establishes new clarity and elevation—a perspective that so attaches to the world that it is not easily distinguished from it. The effect is usually temporary, sustained by something unseen, ultimately to dissolve in turbulence, while that same force brings others.

Among the influences that sustain this process is a faculty of discernment focused upon the impermanence of human creations. This faculty acknowledges that death is embedded within a life process, and is not necessarily the termination of it. Understanding what is perishable and what is not allows one to engage life with a sense of freedom, and to value creative endeavor as a forceful antidote to the fear that accompanies a sense of mortality. Bodies die, as well as theories, relationships, disciplines, institutions and societies. They cannot be preserved effectively without appreciating their frailty, but at some point their preservation becomes trivial by comparison to something else.

If we sanctify that which is not sacred, we find ourselves conscripted to it in a way that disables our ability to take the risks necessary to live a meaningful life. We are moved in that event from living to merely surviving. In order to avoid a fall of this kind, one might ask what happens when sanctity reposes in organized constructs (objects). Where else one might look for it? In the process of answering this question, one might encounter a sense of value that profoundly affects the contexts in which organization occurs, such as the way a principle of social equality functions compatibly with social hierarchy, and the way social harmony functions compatibly with social conflict. It affects one's patience and optimism in the presence of competitive challenges, and the elasticity of thought as order dissolves into rebirth.

Science acknowledges not only the temporary uses of theory, but a tendency of theory to overreach. Once immersed in a theory, one is often motivated to erect a system designed to explain everything. This appears to be the case among a number of scientists who offer genetics as an explanation for the ethical content of our lives—of good. If someone were to suggest that our sense of "good" came from somewhere other than a socially evolved brain, they might be dismissed as unscientific. Their system is in a sense "closed" to spirituality by its theoretical assumptions.

However, systems that claim to define existence ethically may also become closed by organizing experience into a hierarchically structured continuum of development. Within such systems we are apt to find human value labeled and delimited within that continuum—from lowest to highest. The effect is to subordinate value to an order that claims to give value its value. As we have been careful to point out, this puts the cart before the horse. Value confers value on ordered wholes—not the other way around. Theories aside, it is easy to forget that if there is a highest "good" it is likely both ineffable and powerful.

But people are people—it is often said—and know better than to take a given theory too seriously. They draw from value more directly, are open and receptive to mystery, and have much to report about spiritual events. They are also reluctant to concede that the good that animates and gives meaning to their lives doesn't really exist. There are reasons for this.

As mentioned in the third chapter on Moral Freedom, one might attempt to reduce the perception of visually perceived objects to synaptic events in the brain, but that begs the question of what it is that the brain evolved to see. Is it possible that we evolved to experience moral value for a similar reason—i.e., because it exists? Are we being influenced by something that we need to train the mind to apprehend? Is that what one does when one claims to apprehend God?

Perhaps we should leave God out of it. We have already taken notice of the fact that deities proliferate in simple as well as complex societies, and that what we imagine a deity to be is in a strong sense psycho-socially determined. There are certainly magnificent visions circulating about what God is, but to ordinary people, the use of the term "god" is a matter of reference to deity—something that activates the still small voice we think of as conscience. We call it deity because we don't control it, but quite the other way around.

A significant part of the exercise in the preceding chapter on Political Freedom was to leave an opening in the theoretical system being developed, and allow what is unknown to be unknown. Other is not known except in the way we are influenced by it, and the same may be said of deity. We may want to go further now, and consider the substance of these experiences—whether they are material or spiritual. That is quite a problem, actually, because the experiences are, unquestionably mental, but "mental" tied to "brain" is something altogether different than "mental" tied to something that exists with or without a brain. Perhaps it is best to call such a thing "spirit."

Complicating the discussion is that the terms "spirit" and "spirituality" occupy an important seat in human affairs not easily and effectively belittled. There are theories abounding as to what the terms mean. What they seem to have in common is an attempt to go beyond the psychological sense of human identity, through which one defines "who" they are, and attempt to define "what" they are. These theories are, in effect, theologies, some of which emphasize the dependency of mind on matter, and others emphasizing one or more forms of independence.

It should be noticed that materially based theologies are—aside from whether they are true or not—theologies nonetheless. They offer plausible material explanations in neuro-science for phenomena claimed as "spiritual," but they are explanations not about who we are but about what we are. The fact that the explanations are plausible does not in itself establish what we might call "proof" of their truth, but rather that they are in some sense coherent propositions about what they attempt to describe. Whether such explanations are intuitively or ethically satisfying given our experience with the depth and nuance of human affairs, is another matter.

The interesting thing about addressing our use of the term "freedom" philosophically is a tendency to regard boundaries skeptically, inasmuch as they are continually being crossed, a crossing that appears to occur mentally in the experience of value. Thus what we might refer to as a theory of what we are, might be alternatively referred to as a theology. Call it a "liberative theology" because it emphasizes what we might be if in fact boundaries—including material boundaries—are there to be crossed. As such, it is a discussion about allowing, but not requiring, that something called "spirit" exists.

The initial difficulty in addressing this issue is that there appears to be little consensus that there is such a thing as "spirit," and among those who believe that the same exists, a poorly defined consensus as to what it is and whether it is the sort of thing that can be addressed scientifically. Some say that it can't.

This is in fact where the first line is drawn between those who take spirituality seriously and those who do not, i.e., that there is a basic qualitative presence to consciousness that matter does not share. Consciousness is conscious, while matter is not. One can look at one's hands and it is not the matter in the hands that experiences the existence of the hands, but consciousness.

Evidence that a material world exists is that many other persons speak of it as if their experience is similar to ours. This is what science refers to as experimental replication, i.e., that another person under the same set of circumstances will see the same thing The test of whether a flying elephant really exists, for example, is whether everyone else given the opportunity to see it, sees it. Perhaps this only begs the question of whether it is possible for many to share illusions. We are not consciously addressing this question.

As we mentioned in the last chapter, there are shared experiences of deity, which help to persuade those who share the experience that the deity exists separate and apart from their experience of it. If to a group of persons the world appears to be controlled or animated by a transcendent or Supreme Being, they might establish a word such as "spirit" to describe the moving force behind the effect. Then to promote this "spirit" as a self-existent object of worship, they might give it a name.

This was probably more the case for humanity in a more primitive state, where the science of nature was undeveloped, and the world seemed much more like a medium in which intelligent beings, spirits and gods, influence such mysteries as the weather, tides, volcanic activity, the movement and location of game, and human health. Also, it has always seemed that one could, by addressing or orienting oneself to the spirituality of the world, operate within it more gracefully, and at times, change it.

The primitive human was probably more "spiritually" oriented than the modern version, who now explains these mysteries in terms of material laws. Yet, as compelling as a material universe appears to be, it does not change the conclusion that its "material" character is a supposition added to it by consciousness.

There is a developing science of the ecology of survival that notes trade-offs between diet, brain size, social traits, reproductive activity, and so on. The resulting theories of existence emphasize electromagnetic energy, which conscious beings collect from a material world, and use to power electrical synapses that support thoughts and emotions. Admitting all this to be true, the "effect" is still something different than the "cause."

One may call this energy "spirit" so that matter and spirit might at some level coalesce into one substance, but that reduction is unscientific. To imagine a middle region of pre-material and pre-mental "energy" is interesting, but is a fanciful construct offered only to reduce the stress occasioned by the failure to reconcile mind and matter. Is it the purpose of science to reduce stress of this kind, or is science best applied to the ascertainment of fact—here to determine whether mind exists in a meaningful sense independently, and whether mind influences matter?

The eco-systems in which we make our existence is marred by numerous aggressions—the worst of which are committed by human beings. The barbarism and suffering of our species is at times so extreme that it might favor suicide as an appropriate ethical protest to a version of life that unfolds as nothing but a game of survival. But such a protest would, in a manner of speaking, prove the existence of the very thing that the act denied—an ethical presence opposed to a matter based reason to live.

2. A definition of "spirit" that places it apart from some or all of the bonds of matter suggests that "spirituality" pertains to liberation of consciousness. There is a problem here associated with the phrase "spiritual freedom" which, when contemplated, tends to illuminate freedom in its entirety. In the process of examining this problem "spirit" undergoes a shift of emphasis.

The problem is if "spirit" signifies a separation of sorts from "matter," then the phrase "spiritual freedom" is redundant. In other words, to accept that matter constitutes a baseline or principle reality of sorts is to view spirit as an exception to or avoidance of that reality. The most common contemporary view of "spirit" involves the acceptance of matter as a normal reality and spirit as an aberration of that norm. In this view, spirit is the state of freedom from or avoidance of matter.

From this point of view, the reality of matter is accepted, and "spirituality" is a catchword in which all phenomenon of experience occurring contrary to or independent of that reality are grouped. There is little to distinguish or grade the quality of these experiences except that they promote a sense of confidence either in the independence of consciousness from matter, the existence of a consciousness governing the universe, or a relation of sorts between those two spiritual suppositions.

Counted among a number of popular religious orientations is the view that there is a patriach named God who created a material universe and set humanity inside it. This God, while perhaps omnipotent, has done this for no other apparent reason than to amuse himself. After a period of time a human being is removed from matter, an event signified by death. Under this design, freedom is about suffering through matter, and finally being released from matter. One learns to ignore the impositions of that world with the hope for delivery from it. Some such religions claim that redemption is uncertain, i.e., that many are simply not fortunate, smart enough, or moral enough to receive the final spiritual reward that God offers.

This view of God forces one into debilitating rationalizations for the existence of evil. We are forced either to deny God's goodness, or God's power. If it is a male authoritarian God we are thinking of, for example, then he has created a universe in which he has herded innocents into a coliseum, of sorts, to be preyed upon by a Devil. If he did not intend for this to happen then he is not omnipotent. If he did, then despite occasional kindness visited upon individuals within the multitude, he is not benevolent. The appearance of evil, and the personal discomforts arising from it, strongly suggest to many that it is better to renounce such a God on ethical grounds than to believe in a god that doesn't like us.

Given the evidence that supports human dependency on matter, a theory that proposes that matter is only a state of mind would, in effect, be setting the mind of God apart from the human such that God's thoughts are, for all practical purposes, indistinguishable from matter. The obduracy of reality—its constancy in the face of hopes that it be otherwise—is a statement about a God who is a creator of matter and indifferent to the human species. It is quite possible therefore to believe in God, and to live a-spiritually.

To put this another way, the point that defines "matter" is not the separation of human mind from the reality of its physical surroundings, but the separation of mind from God. One can believe in a God or gods, and theologically entertain a multitude of objectifications for God—as an activating principle for all that really is—but one can hardly call the existence emerging from a given theology "spiritual" without the belief that humanity is in some sense embraced and nurtured within that principle. Matter is a world organized against mind—with or without the supposed existence of God. A theory of spirituality might do best by discarding an individuated God.

It is therefore not surprising that the great theological systems of the world attempt to define God as a meaningful ground or basis of an existence that in some sense achieves a state of liberation from matter—not a fleeting or temporary liberation, but an elevated and enduring state of awareness that is undisturbed by matter. God is in that sense an extension, or extrapolation, from an experience of consciousness that is essentially spiritual and is able to prevail against an imagined reality organized against our best interests.

The individuation of God as a transpersonal subject may well be a psycho-logic imposition deriving from experience with value that is transpersonal in nature. This is due to the fact that it is impossible within the domain of human consciousness to imagine a value (state of interest) without a subject as its point of reference. In a liberative theology, God as supposition tends to limit rather than allow that sense of "subject" to be what it is, including something we never imagined it to be. God is a useful, though inessential, form in a spiritualized version of reality, a way of reinforcing good by providing it with an authoritative point of reference.

As pointed out in the last chapter of this book, the experience of deity is to a great extent an imaginary theoretical construct designed to mark an inner terminus on a theory of language. Deity is to identity as other is to object, and by introducing deity imaginatively into a theory of language, we support a vision or theory of language where language communicates as much about who we are as what we want and what we know. Deity is a state of receptivity to a source of influence reposed internally just as other is a state of receptivity to a source of influence reposed externally. "Internal" and "external" are in that sense inventions—suppositions added to conscious experience.

Mystical theologies which describe spirituality as a state of liberation from matter appear to share the view that the "sense" of receptivity reflected in deity and other are states of limitation. Object is experienced through an aperture of supposed access to other, and identity is experienced through an aperture of supposed access to deity. The reality of both deity and other consist in the fact that we can't control them through imaginative manipulation, and they seem to be constantly interrupting us.

These same theologies further emphasize that the size of these apertures constrain us in the spiritual discernment of reality—something that exists on the other "side" of conscious experience. Internal reality is revealed by widening of the sense we associate with deity. External reality is revealed by widening the sense we associate with other. The result of this progression is the realization that the internal and external are one—which is to say, all is spiritual, and matter (separation from good) is a temporary illusion.

This, it seems, is a metaphysical orientation through which one escapes the rejection of God on ethical grounds—which is to view spirit as primary and dominant and matter as secondary and dependent. If one transfers the primary seat of reality to spirit then matter assumes the status of an aberrant or idiosyncratic event. Matter is an act of mental departure from spirit, a relative spiritual state eclipsed by erroneous opinions about God's dominion and benevolence. Spiritually based theologies may in this way appropriate material theologies—where material theories about existence represent relative, not absolute points of reference.

Under this formulation, all is spiritual, but consciousness manifests varying degrees of spiritual awareness. One such awareness may then be more or less free than another. Spiritual freedom presents, in this form, a continuum of spiritual states, the lesser of which labor under the illusion that the world is separated from spirit. The metaphysics of this position is to some extent unassailable conceptually, so long as a material theory is regarded as naught but a way of attributing coherency to illusion. Whether this is a satisfying theology depends on whether one can meaningfully talk about progress along a spiritual as opposed to a material continuum. That becomes, in the end, a matter of demonstration.

However, this view of spirituality is not widely accepted. Most human beings interact cautiously with a world that is believed to hold many physical dangers. This world is, accordingly, navigated through physical inventions and protections. Material recourse comes first, and other options typically arise in desperation after the failure of matter. Spiritual recourse—if taken—normally follows after the failure of material measures. This habit reinforces the view that spirit is a state of exception to matter.

As previously observed, much of what we think of as modern science is committed to the view that the spirituality alluded to by the earth's great religions is nonsense. The controversy is not, to be plain, about whether there is a conscious God entity—like a person with a name—existing somewhere. None of the mystical faiths promoting spiritual liberation adopt any such notion, and in fact denounce it as a form of anthropomorphism. Their preference, in fact, is to replace the name "God" with "good," a state of being that is conscious in a sense that we can't even imagine. The real controversy centers on whether good is something that grew out of matter.

For this reason, there is quite a lot of scientific reflection focused on the use of the term "energy," the appearance of organization in non-conscious structures (revealed in the dissipation of energy), the structure of the brain as a communications network activated by electrical synapses, and the diligent efforts of computer scientists to write programs in machines that can think as artificial intelligences.

An attempt to endorse one view over another here may rightly be here regarded as overly ambitious. There doesn't seem to be a meaningful point to make in that respect. The better course is to examine the ramifications of a more spiritualized view of the universe—perhaps only to allow rather than require it. That is all freedom seems to offer in any event—i.e., some sense of choice of whether to be or not.

3. Why not then begin with a theory of mind by which mind is equated with the gathering and dissipation of energy? It seems that "energy" is in the offing as one of two substrates through which information emerges from and processes through matter. The processing of energy occurs in a computational medium that is essentially binary in nature, and thus accounts at least partially for the materiality of consciousness. It receives stimuli as forms of energy, and translates them into synaptic complexes fueled with energy stored by a living organism. There is therefore a plausible link between features of mind and digitally formed material processes.

The problem with this model of mind is that while it fastens to a viable physical metaphor, transferable between brain and machine, using "data" as a material substrate, there is no comparable substrate for value. However, it is quite impossible—even after much contemplation—to support an experience of the world that excludes value.

There may, in fact be material tissues that enable the experience of value (and thus of "experience" as we define it), but they are not one and the same as value. The attempt, moreover, to equate "experience" and "energy" doesn't solve the problem, even if energy were in some sense linked to value. We would still have to see value as a transformation upon energy—from a physically to a mentally defined energy. We thus have anything but a material image of mind to work with.

Brain I (master)   Brain II (servant)

Rather, what we have is a brain attempting to unite two kinds of existence, one recorded in binary codes and the other presenting itself as value, embracing and experiencing them through selection.

While it is possible to imitate selection by introducing data processing filters, the filter itself doesn't select, but serves as an instrumentality of selection by the designer of the filter. Whether the filter is doing a good job of selecting is a matter of the application of value to its functionality.

Binary states in computational mechanisms bear some resemblance to experiential formations within the brain, enough so that we accept that the brain is to an extent like a computer. But valuative states in mind bear no resemblance at all to programmed filtrations of data used to simulate what we refer to as "intelligence." At a very basic level, we know the difference between a value that selects and a filter that sifts—even though the effects are often similar. The filter can't say if it is working well or not, but a value can. This appears to represent a hole in a material theory of mind—not a small hole, but vast, both in its presence and in its implications.

Whether brain is the sole means of access to value or one of several means of access is not the subject under consideration in this discussion. The point of attention here is whether the quality accessed differs in a meaningful sense from the means of access. A material theology supposes that there is no meaningful difference, while a spiritual theology supposes that there is.

The difference between an information filter, and an act of selection is nicely revealed in the contemplation of scientific freedom. If we were to take as granted that a scientific theory is a data processing filter (quite an oversimplification), one might ask what kind of filter prompted some of the more striking breakthroughs of science. In order for that to occur, a filter that we didn't know about, or for many thousands of years, suddenly activated. However, the major interventions we think of as "science" reflect or manifest a value that is intrinsic to what we are, and rather than emerging from dust, is a stage in the discovery of what we are.

It seems, in other words, that these breakthroughs had much to do with transformations in the value placed on the relationship between human beings and creation, and a sense of awe and entitlement that came with it. One might labor endlessly toward the description of a hierarchy of programs that produced them, but that reduction is itself unsatisfying, and more difficult by far than supposing a dimension to "reality" that is ethical in nature. As a matter of explanation, it is simpler than a mechanical theory depicting mind as an assembly of filters.

As a simple matter of observation, we live in a world where people are moved to tears and engage in heroic risk. To say all this emerged from matter may have some truth to it, but it is also disappointingly superficial, a bit like taking a violin and using it exclusively as a percussion instrument.

The inward movement that is often referred to as "spiritual," the experience of which is associated with very strong emotions, suggests that the ethical content of "hierarchy" is less a matter of altitude than of centricity. People tend to know, instinctively, to touch gently upon subjects that lie close to the center of another's beliefs, and that disruptive contact of that nature can be dangerous. This awareness—of getting closer or further from center—is indispensable to the design of social systems required to move individuals in and out of commitments. As we move inward in our social relationships—from recommending to others what they should do, to what they should like, to what they should be—we know intuitively that we are upping the stakes. We are moving through an inner hierarchy that exists in a relationship with social organization. One thing that emerges from this process—in its social application—is a sense of the sacred.

There have been many past attempts to identify something from which an ethical universe emerges. It is deemed so essential to the nature of our constructed realities that it is accompanied by a sense of embarrassment in giving it a name. The Hindu tradition poses a distinction between an infinite "ground" or Brahman spoken of in terms of attributes of "good" and "meaning," and the deities that emerge from it, while the Buddhist traditions don't speak at all of God except as "being." The Abrahamic religions—Judaism, Christianity and Islam—name it reluctantly, as if existence received its name on account of rather than as the effect of value. Take the first five commands of Moses—here since the dawn of civilization:

You shall have no other gods. (You exist in a direct and intimate relation to the creative power and substance of all that exists.)

You shall make no graven images. (That power does not consist in objects but in mind and in the value it manifests.)

You shall not take the name of God in vain. (You should treat your relation to God as important, as though your life depended on it.)

Remember the Sabbath. (Don't ever forget how important this is, more important than the work you do.)

Honor your father and your mother. (You must respect the circumstances into which you were born.)

The last, "Honor your father and mother," at first appears to be a social imperative emphasizing the authorities that bond the family unit. But this interpretation is superficial, and out of character given the spiritual resonance of the other commands.

Spiritually interpreted as a ground for the existence of "mind" one might emphasize that honoring of parents has more to do with the value of looking beyond the limiting circumstances—mostly inherited—of one's life, and staying focused on the value of one's identity as a spiritual entity, deriving meaning from spirit as ground, or God. Life is a challenging experience, and challenges at birth must be embraced.

This interpretation appears to be what activated the Christian version of spirituality, and the mythic narrative of Jesus, who was committed to the idea that God was his father and that spiritual freedom was a matter of becoming reborn to that reality. As the narrative goes, he was tempted to renounce his calling, but overcame a number of challenges set before him, including a humiliating execution by petty municipal politicians. The story has become a matter of celebration in billions of lives.

The story of this event, whether or not it describes something that actually happened, emerged from ethical traditions that emphasized kindness as a basis for existential transformation, and the assignment of value to lives fraught with institutionally organized oppression. The fact that we believe that such things can be endured, and that we might, as a consequence cross-over into a reality that is in a way imperishable, can be liberating.

4. As an historical artifact, the narrative hangs like a garment that has been stretched to fit many different sizes, and is popular among evangelic religious organizations that prefer to view Jesus as a god and humanity as a pitiful buffoon—by comparison. A figure such as Jesus might be better received today by scientists. They share an expectancy that "truth" in any form is apt to be simple and beautiful, and are inclined to address a diverse and challenging social world without fear or condemnation.

Most interpretations of this narrative, and generally, of scientific indifference to religious figures of old is due to the fact that their lives were recorded in stories that by modern standards are not trustworthy. Scripture is uninteresting absent a prompt that resonates in one's present experience. Even then, the dramatic examples of advanced consciousness mentioned in scripture tend to make "freedom" inaccessible. What is spiritual freedom, in other words, to one who has not achieved rights of ascension, as a Christ or Buddha?

These narratives emphasize the existence of a disparity between an ethical ideal attaching to human existence, and the brutish state into which the human organism is born. This appears as a dualism asserted between an ethical constant in a state of interaction with a constraining environment (both social and physical). The theologies claiming spiritual liberation insist that this disparity is an illusion wrought by human limitation, and that the human organism—as an intervention, of sorts—cleaves the natural complementarity of an inner and an external reality in a manner that produces the image of conflict between them. Freedom is thus defined as a process through which that separation is rectified.

There are a number of principles that companion with this idea, and make strategic use of religious narrative emphasizing human beings who have navigated this sense of disparity successfully. Some of these are as follows.

• The purpose of a given life is to examine and overcome the reality attributed to perceived differences between an inwardly experienced ethical value (good) and its externally experienced environment. The experience of a life acquires meaning through this process.

• An event that overcomes this sensation of difference occurs through the intervention of an elevated sense of value brought to bear on perceived hardship. There is a relationship between this value and the appearance of reality manifest in the transformative impact of one on the other.

• The value is bound inextricably to the substance of man as an entity, and as an individual, and as such is infinitely present, comprehensive, and available. It exceeds all boundaries, inasmuch as all boundaries form within it, and is accessed individually by acknowledging its power, and yielding to it.

These principles, and perhaps others, appear to be central to a liberative theology. There are principles, however, that appear to be more strategic in importance, in that they are aimed toward the activation of transformation. One such strategy was evident in the last chapter on Political Freedom, in an attempt to devise a theory of language drawing a map of mind in such a way that it revealed a bivalent universe drawn asunder by consciousness itself. The map makes it easier to imagine a source of influence both internal and external to subjective experience—still just a theory but a theory that draws the imagination toward an alternative.

There are strategic orientations to spirituality that are less manipulative, and more affirmational. The substance of what one might call "requisite" in the description of a human organization is part of a similar current in thought. Theories of bureaucratic organization once vacillated over an ambiguity of principle, some emphasizing coercive motivation—with suffering—and others emphasizing permissive and consensus based systems—at risk of coordinative failure. With a deeper understanding of human capability we learned that in a "true" organization, efficiency and decency coexist as a single reality. The result allows one to understand health, i.e., that it never was and never could be separate and apart from ethical principle. This, as an expectation subject to demonstration, can have a remarkable effect on managerial temperament, also referred to as "leadership."

Not all such strategies are scientific, of course. One of the more ancient strategies involves an attempt to become dislodged from a state of focus and attention on other. One might enter into a trance, the goal of which is to eliminate the noise and distraction of surface experiences. This process is often referred to as "meditation." Another strategy is to go directly to the inner experience through ethical affirmation of God's goodness, and make petitions to it. That process is often referred to as "prayer."

One of the interesting things about prayer is that it is difficult to do without adopting another strategic position—i.e., the personification of Spirit as one who might be conversed with privately. There are a number of ways to go about this. One might imagine deities or angels circulating around the universe. There are many sincerely offered accounts by individuals in distress receiving assistance in one form or another from such beings, and it is difficult to say if they are delusions or not, especially where in other respects such individuals seem to live sane and productive lives.

Prayer does, however, seems to represent a more circumstantial adaptation to the fact that God—if he or she exists—is too big to talk to directly, and must be imagined as a friend or parent. The illusion falters as one comes to comprehend that if such a conversation takes place, it is not the kind of conversation had with another person, but one going on simultaneously with an unimaginatively large number of individuals at once. But if one doesn't imagine a personalized conversation, it isn't really prayer anymore, but more like asking gravity to be gravity.

It appears, rather, to serve or function as a personal adjustment to a sense of estrangement from good. The more liberated approach toward divine petition struggles less with the personhood of God, and proceeds more as a meditation on the nature of good itself and that good and reality exist in a relationship, such that holding it in one's thought tends to dissolve one's sense of estrangement from it.

This is a point of interest available to someone willing to contemplate the gradients of spirituality with which one may engage. It can be enigmatic and obtuse, or quite vivid. The point that appears near the end of such contemplation is that freedom cannot, by definition, function to belittle or shame a given sense of estrangement, or a given strategic recourse.

In this sense, a spiritual theology tends to be less closed than theologies representing fundamentalist assertions of right and wrong. As we have noted previously, materially based theologies occasionally exhibit their own fundamentalisms, and shift depending on the clarity emerging from an array of divergent theoretic platforms.

5. We have acknowledged that a theory of language is necessary to the understanding of the political nature of the human organism. Does the same theory assist in making a transition between a political and a spiritual view of human experience?

A political view of humanity tends to close to the extent we think of the emergence of deity solipsistically, or rather through the portal of a single individual. The political system functions as an enclosure out of which an ethical reality emerges within a solitary mind as a psycho-social effect. This way of looking at the nature of consciousness opens, however, by examining the real content of other, and discovering that other is not other, at least in the world that other represents.

As one steps into the world of other, it is at once not other at all, but quite the opposite. This is apparent in the interventions we bring to others and others bring to us. It is not merely about sharing information about a material world, but the communications come packaged with moral judgments, and theories about what make a person whole and complete, there with a deity (or many deities) in the face of our own. Try to blend with their world, and immediately one discovers that they are being vetted and examined, and moved toward compliance.

The realization that comes with this and other socially challenging environments is that other is not a very stable notion, but rather an assortment of creation generating organisms—some relatively passive and others not. We receive their messages in the first instance as digital images to be processed by Brain II (see above), but they are loaded with things that act like viruses aimed at the appropriation of Brain I—also known as "me." It isn't such a bad thing, actually, because the process of appropriation is usually accompanied by an array of practical supports. There is also some comfort in the realization that these impositions are not digitalized viruses, but derived values attempting to share experience with our own.

The situation is different when one comes to other not intending to blend, but to change. There it is not about the peaceful co-existence of cultures (political freedom) but with cultural transformation imposed by one on another. One cannot effectively assess this situation within a political model of mind and/or society. In other words, we are obligated in some situations to look past political freedom. Certain kinds of social conflict—while not defining spiritual freedom—help to reveal what it is (or may be). Resolution of such conflict includes an opportunity to recognize the limiting nature of deity, and to acknowledge an ethical continuum on which deity is positioned. In the process one asks whether that continuum is spiritual or material, and thus engages what we refer to here as "spiritual freedom."

The "may be" is an important consideration, because almost all that come to the table of discourse about what spiritual freedom is—this author included—readily concede that they have not achieved it, along with other desirable character traits we associate with ascended beings. Among these traits are "charity" and "humility." It seems, though, that a theology of spiritual freedom might include the following.

• A lens that includes and clarifies other types of freedom discussed in this book.

• A view that reveals limits inherent in a version of consciousness that is essentially political.

• A platform having reflexive stability—i.e., one that does not easily degrade or regress into a political theory.

• A value that supports focused and productive action on matters of real importance to ordinary people.

These requirements might be satisfied with a modification of the images we associated with political freedom. There was drawn a map of mind, the mind of one working to fashion identity within the socially integrative instrumentality of language. The map moves attention through a hierarchy, though not from lower and higher, but from outer to inner, and a state of attention that moves about within this in three tiered language events.

$$+\leftarrow(\rightarrow * \leftarrow)\rightarrow+$$

We have posited this in order to emphasize that being conscious is in some respects a consolidative process that never catches up with itself. It moves like a wave, theory following after theory. From a spiritual perspective, however, the process is also progress. But progress toward what?

An attempt to imagine what that is begins and ends with the identification of a value that embraces the entirety of the organism subject to observation. This proceeds, as method, in a way explained in Chapter 2, where an act of distinction essentially selects an object from a void through the intervention of value. In this instance, we attempt to imagine exactly what sort of value might embrace the experience identified in the foregoing map of a language suffused mind. With that done, most of the work for the development of a liberative theology is done. But there is quite a difference between a theology, and its demonstration, and that is something that should be evident in the theology itself.

If it helps to draw pictures, look at the transformation of subject reflected in a diagram of language to object—from an entity within which value circulates to a value within which language circulates. Call this the "objectification" of consciousness. Thus,

$$\text{``}+\leftarrow(\rightarrow * \leftarrow)\rightarrow+\text{''} \qquad \text{changes to} \qquad \text{``.''}$$

allowed by the intervention of interest "( )" such that agreement between interest and object converts "." as a neutral entity to something meaningful, (+). We have, in doing this, moved "+" as a placeholder for other into a state of intimacy with value that intervenes (selects) the mind evident in an individualized consciousness, and captures it (so to speak) in a given moment. The moment could be a state of mind, a lifetime, or a series thereof.

On Freedom: Spiritual Freedom

Within this frame of reference, time appears to consist in the transformation of "." to "+" within value ( ). Time records progress from a state devoid of meaning to one in agreement with the interest that embraces it. It is here one and the same with the meaning one attaches to the idea we refer to generically as "man." This idea can mean any number of things. It can refer to the human organism, or it can refer to something we imagine that the human organism is becoming.

One can allow within this frame that a value of this nature is unbounded, only because one allows that "boundary"—as an idea—occurs within it. So, it may be meaningful to attach a symbol to the point of observation from which this value extends. In this theology, call that point of observation "infinite" or "*", not because there is a concept of infinite that can be here described, but because we acknowledge a state that precedes and thus exceeds the assertion of boundary. In so doing we have identified a state of agreement that is equal in kind—though not in quantity—to this state.

(+)

Importantly, it is a matter that can in a meaningful sense be shared, even though the manifestation of it is individualized in the being and experience of a given person, and thus support a diagram as follows.

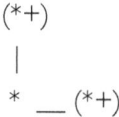

Other has shifted from a position exterior to consciousness to one of shared occupancy of an infinitely dimensioned universe, each person representing a world entire. If one chooses to equate the placeholder "*" with God, then this diagram might translate into a statement to the effect that (*+) represents God's understanding of what man is. This is occasionally referred to as the "Atman," "Buddha" or "Christ."

Important to the understanding of this theology, however, is that it doesn't require the ascription of content to "*" other than the fact that it precedes or subsumes boundary. It doesn't require the name "God" as an attachment to it. The value extending from it doesn't have to be any particular value or emotional state. Such assignments are allowed, but not required.

It could hardly be called a theology of "freedom" if it did not allow for its own discovery, if it failed to supply a basis for individualized acceptance and affirmation, or if it could not be freely rejected. This could be the case if one discovers, for example, that the affirmation of one or more gradients of the value "love" produces meaning and satisfaction in the experience of a life, and produces transformative experiences and the healing of estrangements. One is not required to live in a spiritual universe, but one may.

One might, in other words, conclude that love is not a transcendental value, and that it is naught but a biological adaptation that promotes cooperative behaviors—such as a hybrid of emotions fusing affection and social responsibility. One might conclude that the testimonies of others who speak of spiritual experiences are delusional or dishonest. In the process, it is useful to remember that honesty and sanity are themselves included in the subject material of spiritual freedom.

That such choices occur within a theology welcomes one to examine and compare theologies—theories about the real nature of existence. The practical importance of doing so is that it affects the kind of social reality one will attempt to create in a state of partnership with other, as well as the extent to which one is likely to acquiesce to social disruptions.

6. A fair measure of the content of spiritual freedom has been examined already. In what we have referred to as "intellectual" and "moral" freedom, there is a process of discovery that moves from more to less restrictive experiences, and the correlative effects of more effective navigation of a complex environmental surround. In the case of intellectual freedom, it involves the navigation of objects through the insertion of interests. In the case of moral freedom, it involves the navigation of interest through the insertion of selfhood.

Take the case of intellectual freedom. There the movement from constraint to liberation involves a series of interventions of value within a cycle that is itself activated by value. What we refer to as a "logical process" is, as a matter of structure, the effect of the integration of one and only one intention. The change that allows this integration is identified as an independent variable called "capacity" enabled by the conversion of deliberate processes into automatic processes—such that intending or trying is either unnecessary or has in some sense vanished from the intellectual landscape. Each and every process—such as securing, intimating and ordering—engage the intervention of new interests that derive their value from an interest regarded as primary. The primary interest—also referred to in organizational literature as "declarative"—marks both the emergence of a new order of complexity and the completion of an old order.

Each qualitative shift of logic—from phase to phase within an order, and from order to order—involves adherence to the form described in a liberative theology. The form $1(1+)$ marking entry into an order follows or is subsumed by $*(*+)$. Boundary occurs within interest, not around it. This solves the Russell paradox.

This being stated, one might consider a position that is in some sense anti-theological. That is to say, the identification of a form $*(*+)$ is a matter of extrapolation or projection of a material process into a spiritual process that is merely supposed. Under this formulation, theological phenomena are best understood as a matter of reflection, or rather, thought thinking about itself. This is to say that our supposition about the nature of thought itself is erroneous—i.e., interest occurs within boundary, not around it. This engages the Russell paradox.

This latter version begs the question, in a manner of speaking, to the point of infinite regression. What does it mean for thought to think about itself? To do so it must have an interest in itself, and must direct interest from a point of observation. What is that interest, and what is that point of observation?

If there is no interest supporting self-reflection, then such reflection comes to appear as a hall of mirrors in a carnival, and such is the way philosophy appears in many of its esoteric derivatives—engaging a solipsistic view riven by differences without difference. If the process is meaningful, it is still necessary to ask what gives it meaning, in which case we are moving the human as observer toward a value that regards human existence as a matter of importance.

We examined this value in the discussion of scientific freedom—where it seems that science itself was enabled in significant part by a sense of expectation or entitlement to know the cosmos. It is likewise implied in the concept of maturation that moves from one order of complexity toward progressively more meaningful and inclusive enterprises. It is evident in the movement from an isolated moral universe to one that takes on the needs of others. That some call this process "love" is a way of giving meaning to what might otherwise mean nothing.

In the process of achieving moral freedom, human beings take an inventory of desires presumably activated by a very long process of genetic reproduction and place them on a table of examination within a different kind of interest. As a consequence they change, and as a result of a process of change, selfhood moves from potential existence to meaningful existence.

In a moral universe, actions that are counter-intuitive from a strictly biological platform become compelling assertions of existence that vary from biological rationale. People are inspired by and sacrifice themselves to principles that are deemed indispensable to it, and are embedded in the stories they tell themselves and others about their lives.

But even these generalized versions of selfhood, that come alive in the form of tribal deities, somehow get moved to the table of examination, so that societies can form and communicate in close proximity to one another. The human organism has, within the last five hundred years, devised governmental structures that protect against social destruction likely to emerge in the form of tribal rivalries. Doing so occurs within a perceived interest in preserving and sanctifying individualized expression of selfhood.

The innovation was deliberate, and inspired by the conscious realization that even the gods themselves emerge out of value placed on existence. They are boundaries occurring within value. Remarkably, they appear to set such value apart as ground for existence, or rather, *(*+).

It is not insignificant that people are inspired by the progress of science, by the expansion of domains of complexity, by the achievement of moral satisfaction, and by the formation of just governments. The meaning ascribed to such inspiration draws people to it. As a theology—a theory of spirituality—it appears therefore to take care of itself, and to resist its own dissolution. On the other hand, a theology based on esoteric self-reflection—a valueless mental process merely reflecting upon itself—doesn't do much to take care of itself, and almost by definition, reflexively dissolves.

7. The kind of inspiration that people derive from ethical self-reflection on the meaning of their lives comes out in their political conversations. Their views, rightly or wrongly, are full of meaning about things that ought not to concern them, and that at times are regarded seriously as matters on which their lives depend. Perhaps to oversimplify, the passionate nature of political self-expression reveals, among other things, the existence of spiritual priorities within the landscape of human discourse.

Much of what people pursue as good and decent social objectives, are biologically speaking irrational. The religious value of loving one's enemies—prominent among a number of religious movements emerging from the Axial Age—stands out as a test of sorts of whether one truly embraces the value of human beings. It is, arguably, foundational to the maintenance of peace within a social universe riven by intra-species competition. However, it is not clear at all that we discovered this through evolutionary processes functioning within time spans on the scale of genetic selection. Rather, it appears to be biologically counter-intuitive.

There may be plausible reasons why hatred violates evolutionary law, most of which reflect the practical value of group solidarity and cooperation, but those reasons do not provide an intuitively satisfying explanation for why many have risked self-destruction by extending benevolent regard to vanquished adversaries. Such acts may have resulted in more stable and robust social orders, but the discovery that this was the case occurred through a willingness to test the hypothesis—i.e., whether there is agreement between kindness extended toward an enemy (or a potential enemy) and productive ensuing relationships. In the evaluation of future outcomes, the risk of self-destruction is generally regarded as unacceptable, but such risks are taken. In the final analysis, it appears to be less about contrasts in what is more plausible as a matter of theory than in the ethical comparison of divergent theologies.

There are, similarly, causes brought on behalf of people having no political standing at all. Many are willing to die or be imprisoned over whether it is right for a mother to kill an embryo. Despite the fact that society may be safer upon the death of a criminal, many are passionately committed to legally proscribe death as a penalty for crime. The idiosyncratic nature of such views are, in a sense, more sensible in a spiritual than a biological universe. Depending on one's theology, the same person may feel compelled to preserve the life of a social enemy and end the life of an embryo, due to cares attaching to the meaning of life.

Typical of spiritually motivated individuals is concern toward the suffering of other human beings. But since suffering is frequently accompanied by vile behavior, intervention can be dangerous. That compassionately disposed persons are attracted to this and other forms of suffering, to be healers of it, is an example of what—in a political universe of individuals asserting self-interest competitively—would seem illogical. But to someone interested in proving the value of existence, the dangers one might associate with such interventions are regarded unimportant.

In the present era, there is an unqualified aversion to legally sanctioned ownership of one person by another. The institution of slavery effectively displaces choice that one has over their life, the product of their labor, the composition of their family and the establishment of social respect. The slave is thus another example of those who do not have political standing, like the embryo and the criminal.

Yet throughout history, people have been willing to accept extraordinary biological risks to avoid enslavement, and even to deliver others from that condition. But spiritual value often has difficulty in sorting through the nuances of it. Slavery is difficult to identify in many cases, and may be voluntarily accepted. Servitude, as well as hardships attendant to it, does not necessarily debase one's selfhood and may in fact uplift it. And for a number of reasons, economic inequality is essential to the ship in which society floats. However, despite an organizationally situated belief that involuntary servitude is a necessary evil, it has been cast aside in favor of a different organizational vision.

Ordinary politics is—in a manner of speaking—debunked by spirituality. We can devise political theories for why spiritual people behave charitably toward the helpless, but it is easier, perhaps, to simply admit that consciousness is unconfined to narrow interests and is, by its nature, magnanimous.

The opposite assumption about human nature casts politics as a processing center for self-interest. It supposes that the reason why we are interested in politics is that we hope that by understanding it we might affect or avoid being harmed by politically motivated behavior. Our interest is therefore mainly strategic, inasmuch as we can form a list of things to do that will yield social benefits. But political strategies have limitations, and may lead to disastrous conflict—even strategies of peace that are elegantly conceived in constitutional parchments.

The politically functional world we think of as "ordinary" comes with effective practical rationales that are not designed to address extraordinary disruptions. They do not address within their legal structure, for example, how to handle an enemy who has made a devoted commitment to destroy us.

Assume that there is no means of influence—any words or acts— that will persuade that person to abandon his or her plan. If offered money, they will use it to purchase armaments. If ceded land, they will use it as a position from which to mount an offensive. If given authority, they will use it to organize others with them. If ignored, they will use the time gained to organize their attack more cohesively. In this situation, it appears that violence, while excluded from a political order, is included in the reality of the relationship one has to another.

The situation has no political solution, and the question arises over where one looks to find a sense of obligation that addresses conflicts that are irresolvable. There are, unsurprisingly, a set of options one can choose—i.e., more than one thing to do about it, each carrying its own signature of risk. When confronted with such a dilemma, there is an opportunity to examine what "power" really is. Is it a matter of who wins a contest of wills, or is it a matter of identifying a source of influence that exceeds the boundaries of a given theory of self-interest? The latter of these two ideas explains in significant part the nuances of many of our most important social transformations.

There is, perhaps, a difference between the person who risks persecution in order to promote freedom, and one who does no more than organize resistance to oppression. One seeks to transcend the symbiosis of belief, of action and reaction, while the other rehearses it. One demonstrates power by asserting values that resolve estrangement between persons, the other perpetuates estrangement by attempting to overwhelm the behavior of others with acts of coercion. How does one alter that script?

If there is a point of nondestructive resolution to this conflict it is apt to be a result of one side (or both) taking risks that make no sense at all politically, the value of which have much to do with the fact that one side (or both) managed to refrain from hating the other. Finding a solution can't occur without first looking for one, and may involve having to identify and root out influences (people) interested in promoting a violent outcome. Not all perspectives in a conflict are legitimate, and not all are constructive, but the perspicacity that allows one to navigate such a field, though troubled, is typically regarded as a matter of general celebration.

The celebration comes as more than a sense of relief attaching to the fact that war was avoided, or favorably ended. It appears, additionally, to do with the fact that one may have discovered a way to demonstrate that the extension of grace toward "enemy" can be done in such a way that avoids self-destruction. Such celebration, brought on by ethically situated motivation, is the way that freedom, as a theology, sustains itself.

8. The sense that there is a difference between a regressed and a progressed state of spiritual development is evident, as discussed above, in the fact that we celebrate breakthroughs in the peaceful resolution of ordinary as well as extraordinary human estrangements. It is also evident in the way we evaluate the history of human affairs, and the formation of models that make sense of history.

One might argue that the notion of human progress in the development of science and the formation of industrially scaled economies are the product of a political breakthrough brought within the lens of a liberative theology.

There are opponents of this who argue that the notion of human progress toward enlightened consciousness is grossly overstated, emphasizing numerous atrocities committed within wars brought on by a sense of moral entitlement of one race or culture against another. They emphasize that resource depletion, and pressures resulting from viral plagues and/or climate change will reveal evil of unimaginable magnitude in human affairs in the not too distant future. They further argue that the plausibility of such developments in itself implies that spiritual freedom has little, if any, substance to it. Thus the attachment of "spiritual" to any aspect of the human condition is an extravagant instance of hubris in the self-reflective tendencies of human language.

They may be right. But then again, they may not be. A spiritual theology allows that progress was and remains a matter of choice. Let us then return, momentarily, to the image formed by a theory of spirit that casts humanity in the embrace of the value of existence, referred to by many as "love." Making use of the image drawn above, we can depict how the image of perfection allows for the comparison of matters of "progress."

```
   (*+)                        (*+)
    |      allows to see        |
   * __ (*+)                   * __ (~*+)
```

This diagram replaces to the right one of the signs for "man" as the object or embodiment of value with its negation. This, in a way, restates a principle covered in the chapter on maturational logic, noting that affirmation is the event that allows one to see or appreciate a negation. Perhaps by imagining a "Christ" one does, at least imaginatively, bring a supposed "Antichrist" into view—and an historical eschatology with apocalyptic overtones.

Unsurprisingly, some fundamentalist sects of the Christian religion are preoccupied with an end-time that bears resemblance to what more economically mindful authors identify with the Malthusian disaster. They are likewise united in their claim that the human species is powerless to avert this catastrophe.

This sense of inevitability and powerlessness is where a liberative theology differs. The expression (~*+) can mean "the opposite of

(*+)"—mind opposing, or otherwise represent a preformed state (* minus n +)—becoming (*+). Within this theology there are no certain futures, but there are reasons to take existing progress as a matter of encouragement.

There is, moreover, no Antichrist, real or metaphorical. The underlying reality is spiritual, and departure is a material illusion in the process of rectification. Whether this works as a visual template to understand the status of human affairs, past or future, is a matter of demonstration.

What it suggests, perhaps only as a working hypothesis, is that gloomy forecasts about the future of human civilization are likely the product of a closed system of thought—i.e., that categorical denial of the spirituality of humanity dooms it at the moment environmental and political pressures become too complex to manage. There is no value that is otherwise redemptive.

One way to close the system is to suppose that our model of language is a-spiritual, or rather, that deity identifies a materially emergent property of consciousness. One might, in other words, accept the model it represents, deity being a presumption we make about an "inner" source of influence, but insist that it is a digitalized projection. It's information is something that comes to us without the conscious exercise of control over it, but that each and every participant in language makes it up. Philosophically speaking, it is hard to argue for or against this proposition, except to return to the problem with the use of a digitalized substrate for the emergence of value.

A theology of spiritual freedom, on the other hand, allows, but does not require, that the emergence of deity serves as an indication that the value attributable to existence exists trans-personally. Such a theology is relatively indifferent to claims over whether tissues of the brain are the sole means of access to this value, inasmuch as digitalized constructs form and have meaning within it, and not around it. The issue is whether it exists, while the means of access to it is a secondary consideration.

Of more practical importance is the way we interpret differences between what we imagine to be "spiritual" and what would be relative states of progression towards it. Historically speaking, one can plausibly argue that the emergence of industrial economies is the result of a shift of spiritual awareness.

This subject may better serve as the subject of an entire volume of analysis—a book unto itself—but certain observations can be made in a perfunctory way. Suppose, for instance, that the low relative productivity of preindustrial societies was the consequence of closed and controlling government structures. The innovation we associate with political freedom may be regarded as a breakthrough of value. It held to an optimistic theory about human nature as essentially cooperative and productive. It was implemented at risk that many would regard unreasonable—i.e., strong resistance from centrally positioned rulers, and much doubt over whether social order would collapse under complex and divergent social pressures. The prospect of allowing resource accumulations in independent organizations within free marketplaces had for millenia been regarded unthinkable.

The implementation of a substitute for centrally controlled government had much to do with the belief that the evil was not due to human nature, but to fears related to personal security. The solution, it seemed, was a matter of effectively separating military structures from economic interests, and establishing rules for competition that avoided death or enslavement as a consequence of failure. It promoted a theory to the effect that social justice might promote rather than inhibit economic productivity.

While it was acknowledged that tribally formed deities make political discourse contentious, they are socially adaptive—something apparent only within the setting of an elevated notion of humanity. They can't be killed, but they can be reasoned with. It required some time, effort and risk to learn that this was the case.

Such wisdom fails to achieve focus through a lens that looks upon a human being as a slavish beast, but by contrast, as valuable, and driven to fault by environmental pressures. It may also reflect a religious nature which—while often controlling and obdurate—tends to subordinate to a spiritual nature that progresses. This discernment has two important implications likely to influence the course of human affairs in the future.

The first of these implications involves the preservation and consolidation of benefits that have appeared as a result of certain breakthroughs in the understanding of the human organism. Understanding the reasons for the social innovations that account for social progress is probably essential to the avoidance of a slide backward into disappointing behavioral patterns. If justice is the spiritual motor for modern economies, injustice will likely derail them.

A second implication of the importance of spiritual discernment involves the relationship between pre-modern and postmodern economies. If the success of modern economies consists in the advancement of the value of human existence, the marginalization of the effects of fear, and the accommodation of human limitations, we might approach and address the impoverishment of pre-modern societies with greater urgency. Spiritual freedom stabilizes through progress.

The choices pertaining to the effect of inequalities within and between economies may, in the last analysis, be more a matter of theology, and the adoption of a view of human nature that allows for the wise selection among many available political strategies. The problem of strategic intervention through government is solved through a process of discernment enabled by a spiritual orientation that is ideologically durable and open—not something easily established. But if it were automatic, trying would be unnecessary, and whether someone chooses to try, or not, depends to a great extent on the expectation of gain that precedes it.

The urgency of intervention is practical.

The urgency is also metaphysical.

The lesson of recent history appears to be that practical and metaphysical considerations are joined.

9. Near the middle of the twentieth century, there was a great war. The war emerged from the rubble of a collapsed free market, and of institutions that had supported that market. It was waged with ingeniously devised and mass produced weapons—putting industry against industry, and included an attempt to kill all the members of a major world religion. By the end we had, as a species, learned to destroy whole cities in a single attack, with a single weapon. Was it a necessary war?

Within a liberative theology, there is no simple answer to this question. Some years earlier Mahatma Gandhi began to lead India to independence in a series of demonstrations of the power of nonviolent resistance against oppression. This achievement awakened many to the possibility of a spiritual solution to what appears to be politically irreconcilable conflict. In the case of India, Gandhi knew that England would not wage war against freedom in the consciousness of India's people. Lying remotely beneath the aggression of English imperialism was a conscience likewise activated by freedom

The Germany that faced Europe was not like the England that faced India. England exhibited spiritual deficiencies in its oppression of the people of India, but it was nonetheless committed as a nation to the idea of freedom. When India spoke with a spiritual voice for inclusion there was a place in which the same could be heard. Germany's transgressions bore no resemblance to England's condescension, and were more like maniacal acts of self-preservation. It also appears as though Germany's mania was exploited by well organized and secretive economic collectives with an interest in profiting from slave labor.

To many Germans, the progressing world had failed and could not be trusted. That left Germany alone and vulnerable to change it could not stop. Germany's brittle adaptation to the perceived threat was such a withdrawal from spirit—love—that social theorists will be seeking ways to explain it for centuries.

One might imagine that history is full of errors of judgment over matters of war and peace. Opportunities to construct peace out of conflict are lost. Peaceful overtures are wasted on unworthy opponents. Certainly an important determination before putting forth efforts to make peace with others is whether their perspective allows it, and if so, of what kind of peace it allows. Since it is not about the person, but their relative state of captivity, the spiritual skill applied involves separating persons from their thoughts, so that we may accept one while rejecting the other.

Perhaps the more important lesson derived from a discussion of "progression" is that while there is spiritual unity among individuals within the human species, there is antipathy between various mental states—particularly where spiritual progress challenges static perspectives. The freedom we associate with permissive marketplaces might therefore clash violently with ideologies that evolved to promote and protect the centrally controlled economies we associate with agrarian civilizations. .

It is understandable, therefore, that the coexistence of post-industrial and centrally controlled economies is—in our modern world—somewhat awkward. We nonetheless stare with bewilderment from a post-industrial platform at societies that function within agrarian mentalities—i.e., tending toward totalitarian organization, and rigidly stratified social castes.

To make matters worse, indiscriminate attempts to install liberal political structures tend to destabilize governments operating within a control based ideology. To a feudal monarch, for example, it is difficult to imagine a permissive view of the formation of economic collectives now referred to as independent "corporations." Their formation would soon put the monarch at risk of assassination, a common occurrence in nations attempting conversion to modern political structures. There do, in fact, seem to be quite a few sad modern stories of very decent and popular political leaders driven to oppressive extremes at the sheer difficulty of surviving among rivals.

Free markets supported by a relatively permissive government is therefore a paradox. No government except a strong and stable government can afford to permit free association within a free market. Strength and stability, however, are difficult to sustain while yielding a significant measure of control to organized collectives. There are solutions, of course, but these are meaningless theoretically without devout commitment and an unusual power of discernment.

Whether and to what extent a free society requires a strong government was an issue placed before an entire nation in the middle of the nineteenth century, when civil war emerged between the northern and southern states of America. The North American continent was colonized under two different states of motivation, each representing a different theory about human nature. One was to avoid the interference of European governments with the practice of religious liberty, and the other was to acquire resources, primarily land.

As a result the South evolved in accord with European model of feudal estates granted under monarchic authority. It held to the aristocratic view and purchased labor in the form of slaves brought by force from the continent of Africa. The result in the North was the gradual formation of a more modern industrial economy, fueled by a passionate attachment to a free world of opportunities. This expectancy attracted the immigration of many others, as industrial labor—who were discouraged with the stagnant European feudal economies.

The North, empowered by greater numbers, and by the economic success of a free market, promoted laws that advanced and justified that market, and supported leaders who renounced the institution of slavery. The South, defending its identity, advanced the same rationale that had sustained feudal aristocracies since their inception, i.e., human beings (Africans) are animals that need to be controlled and protected by aristocratic institutions, and would be lost without them. The southern social order was thus regarded sacred, and the North was vilified for its ungodliness and greed. As the North advanced a law emancipating the slaves, a situation for which there was no political solution arose.

From the perspective of the southern states, the solution was a legal solution based on a state's claim of entitlement to self-determination. Within the universe of political ethics, there is coherency to the logic of that argument. If it was the sovereign right of a state to choose, or refuse, to enter the federation of states, then it was the sovereign right of a state to secede from that union, particularly if no harm to other member states would result from that act.

But from a spiritually evolved perspective, it is not that simple. There remains the question of whether withdrawal from government is in other ways disruptive of institutions that are necessary to the continuity of a free political system. The question is comparable to that of surgeons who must ask whether, in the process of removing or repairing an offensive bodily organ, they might kill the patient altogether. Spiritually motivated political interventions thus require an accounting for what to do with the governmental corpse after political and economic theorists have made their point.

Intervention seems different where, as in the conflict between northern and southern states, slavery is endorsed by states over which the federal government claims sovereignty. This challenges a government to decide what it is—either of and for the states or of and for the people of the states. That decision invokes reference to the value of human existence and by implication which interest government serves. In the case of civil war between the states, a federalized government decided that the value of human freedom provided ample reason for it to resist its dissolution so that it could—among other things—address the atrocity of slavery committed within its borders.

The institution of slavery was established within a "free" marketplace, applying the principles of contract and property. A liberated market moves swiftly and powerfully, and will fragment without effective regulation. A failure to protect government after bringing these influences to bear on the world is like starting a fire in the forest without assuming responsibility for its containment. If a government ordained on the principle of freedom cannot preserve itself against secession, then freedom is not more than a theory.

Abraham Lincoln expressed this idea before a gathering of citizens shortly after a battle that decided the outcome of the war. The speech was a prayer to spiritual freedom, which included an acknowledgment that the sacrifice was made so that others might continue and ultimately prove the reality of freedom in a state which, though imperfect, nonetheless invited and enabled progress—an open system. Freedom, once constituted in government, could no more responsibly cede authority to act on such matters than could a mother cede her child to a burglar.

After considering the suffering that a Civil War would entail, one is likely to be moved by the spirituality of many of those who endured it, as they addressed through sacrifice not who they were, but what a human being is. That a national identity could be defined and empowered through such a conflict reminds one that tragic events can produce changes that continue in their importance long after the pain of those events abates.

As the South was attempting to form itself into its own nation, the monarchies of Europe were forming themselves into nations. The sense of separation occasioned by the competition of nations for resources resulted in a war aimed at stemming the aggression of Germany, which had by the beginning of the twentieth century, developed an ideology that claimed a right to territories associated with its ancient cultural heritage. The prevailing European alliance, with the assistance of the United States, defeated and thereafter punished Germany for its transgressions.

It was a temporary victory, and in less than a generation, Germany returned from its defeat, reinforcing its earlier claim to an expanded German territory with a claim of cultural and racial superiority, and a Jewish based conspiracy to weaken and destroy a sacred German identity. Despite the fact that its aggression involved profound suffering, the victors sought reconstruction. The preservation of Germany after its second defeat, along with the success it found in a strengthening market, somehow helped to awaken it from its nightmare.

All comments we would like to make about the triumph of good over evil hardly make up for the many craven acts committed by human against human. We can draw as many despairing as hopeful views about human nature, but can, however, derive some encouragement from past conflicts that our devotion in some sense strengthened us in battle, enhanced the peace that followed and allowed us to identify and suppress the recurrence of similar conflict.

The attachment of finality to a military victory misunderstands the conflict. The conflict did not arise because evil sought to defeat good, but because progress stimulates resistance from those thoughts, values and identities which are displaced by it. Freedom thus responds cautiously to victory, knowing that the struggle was not about persons, but states of mind. It is also useful to acknowledge that states of mind are not as easily defeated as armies, and may reappear in places and personages one would hardly expect.

There are scholars of the American Civil War who say that this war was decided at the battle that prompted Lincoln's address, and that this battle was won in the defense of a strategic mountain top. The soldiers there, having spent all their bullets, made a strategic rush without effective armaments, and chased the Confederates away from their best chance of victory. Within the same battle, but at a different location, another group of Confederate soldiers made a suicidal rush into the most heavily fortified part of the Union line, and the battle, along with the war, was lost.

Scholars of the war against Germany likewise remark on the self-destructive course of the Nazi regime, which at select moments in time held opportunities for victory. While many argue that luck ultimately decided these outcomes, there is something in the details that suggests differently. There are groups of individuals that may have ambitions similar to those who brought forth the Third Reich, but they are apt to discover that it is human spirituality—as opposed to human psychology—that presents the largest obstacle to their doing so.

10. As much of the world rejoiced over the containment of the predations of an aggressive German nationalism, the celebration appears to have neglected part of the progress necessary to avoid the violence that will accompany the future efforts to resist spiritual progress. As one speaks triumphantly of the power of a free market economy, they are likely not to admit that it is only evidence of freedom, and not its equivalent.

Unless a free government places limits on the predatory activity of such an economy, the collectives that form within will, for competitive reasons, increase in size, and may upon achieving a certain magnitude, secede from the principle of organization under which a free market is established. This occurs as economic collectives strive to avoid or emasculate the laws which a strong government would enforce to insure fairness. Without these competition degrades, revealing the difference between the world of material selection inhabited by animals, and a world of opportunity serving humanity.

This tendency is particularly dangerous where these collectives seek to expropriate the governments of foreign markets. To a degree, this theft is assisted by the nations from which these collectives emerge. There is very little to constrain exploitative behavior, and even though there is advocacy of a free market under the regulation of a free political system, the developed nation cannot under these circumstances establish credibility. To spiritual freedom, there is probably no greater blunder than to attempt to impose a transformative change upon another without having first established trust.

For a number of reasons, it is impractical to prescribe a rule or recipe that would end violence among persons and nations. As one takes up a cause, or the cause of another, the act is apt to stimulate opposition that invokes condemnation of the actor. This causes one to reflect on whether good deeds are especially reserved for those who have achieved perfection, or whether there is a way to help others grow while needing growth. Some reassurance comes from the fact that risks associated with undertakings of this nature are considerably more manageable where cultural hubris is effectively displaced by a spiritually situated humility and optimism.

It is often said that "trials are proof of God's care." Whether this is true or not, it is probably a good rule of conduct for one to refrain if possible from the presentation of adversity to others. It would require a certain omniscient vision into the depth and complexity of another's predicament, as well as the power to resurrect them if they should fail. Trials are thus something one does against another, and not on their behalf.

The spiritually evolved consciousness promotes evangelism reluctantly. It is known—or rather assumed—that resistance will follow. Usually the most salutary influence one can have is through example. Otherwise, one's role in their encounter with others, as part of the demonstration of spiritual freedom, is to compassionately assist them with their adversities. This assistance adds much to one's experience of life, but the risks and benefits of this may be enhanced if certain principles are valued. These recognize and respect the fact that awakening is a progression, and is only partially complete. Some of these are as follows.

1. One should not help another without their permission.
2. Meet them where they are, and not where they should be.
3. Avoid condemnation of the person, and address only the error of their thought.
4. While addressing the errors of others, we must address our own.
5. Endeavor to allow rather than prescribe Truth.

Truth is what it is. It does not appear and disappear because a person or a group of persons believes it or doubts it. It does not become truer because a theory or theories bring one closer to it or places thinking in alignment with it.

The errors we embrace, while seeming true, are the product of our limitations. They are comprehensible and fathomable. We have power over them because we made them, and yet they do, for a period of time, blur and obfuscate receptivity to the truth. An encounter with a predicament that suggests grave, irreparable and unavoidable harm may be met with the realization that it is not more than a theory of harm, a mental packing crate the limitations of which allow only a few possibilities.

These principles, by which favorable transformations may continue in and among persons, or in and among cultures, may assist one in practicing discernment in the exercise of evangelism. There is a measure of gratification, however, in allowing that none are excluded from freedom, and that a new sense of existence can mark an improvement upon those that precede it.

While it may be true that the perfect achievement of spiritual freedom includes the experience of a value that negates any sense of penalty associated with a material world, the greater question is what spiritual freedom means to those in transition, i.e., in the process of redefining right and wrong, pain and pleasure and reward and punishment. In some respects it is more remarkable to witness kindness given at great expense by one who does not believe in any kind of god, than to witness the same in one to whom God has spoken on the mountain.

If we ask "who does God love more?" we may be asking the same as "who do you love more?" By its nature, spiritual freedom has difficulty with this question.

Postscript

Some time has passed since I was first willing to endorse this work as complete, and to summarize its contents in the form of prefatory remarks. Upon my most recent attempt to address the issue of "spirit," I am drawn back to a comment I made in the first chapter on "Scientific Freedom." The comment was that the nature of this inquiry is biased because the treatment of freedom as something real embraces a value that influences the form and content of statements one makes about it. Ultimately, this value is grasped—though fleetingly—in terms of a universe that allows rather than compels it.

The contemplation of this idea helps to reveal the tendency of other theories, some spiritual and some not, to overstate themselves, and the realization that they are promoting prescriptive theologies—and quite a sense of obligation to go with it. They congest rather than clarify our sense of obligation, and thus become like brambles. Theories becoming theologies can feel limiting, particularly when formed within market driven academic disciplines each advancing as if on an appropriative mission. As a consequence, they don't seem to be communicating effectively with each other.

It therefore occurred to me that a theory ought to speak softly about itself, while shouting to some extent in its applications. Like mixing a color to paint with on a blank sheet of paper. Be careful what you mix with a color, for you are apt to end up with gray. Since there is plenty of gray out there already, the moderate and effective use of color is to apply it as an opportunist and with some discrimination. In that way, one might find boisterous expressions, without being as observer either boisterous or reticent.

And so it is with freedom. It is an opportunistic doctrine. It doesn't promote the elimination of constraints, nor does its theology suppose that resistance is unreal or meaningless. It is very much the opposite of that, both as a theory and in its application. It reveals itself through the existence of boundaries, and a process consisting not so much in blasting them all away, but by handling effectively the one standing directly before. In like manner, as one confronts a material theology, perhaps the best response is to say very little, except to ask—where does "love" fit into all of this?

Most of the time, and in most situations, it doesn't seem to matter whether it does or doesn't, but that fact neither avoids nor diminishes it. Freedom becomes important in the face of incompatible alternatives, as in whether to go to war and what it means to win. There is a great deal to say about the strategies that bring people together, and the rivalries that drive them apart, that were not addressed in this book, and are best reserved for another.

Human nature figures prominently into the formulation of solutions to problems relating to the need to extract a living from the world. The most innovative solutions respond convincingly to the need to do so justly, and the expectation that there is no real trade to be made between social justice and social productivity. This principle has yet to be tested under the stress of scarcity in the generation and management of energy. The test is approaching.

Perhaps the solution will come with the realization that we make economies, they do not make us, and that our biological drives only partially describe what we bring to the table of dialogue. We may be closer than we think to solving the problem, but equally, be closer than we might imagine to some very disappointing outcomes as well. Are we as prepared to apply our wits to a science of human organization as we are to physical science? We might choose to resolve our problems in a few years, but it may just as easily be one thousand years.

I have placed much importance on the research conducted within the last fifty years revealing constants in human maturation. The developmental courses outlined by Piaget, Kohlberg, and others have been revealing in their own right, but their importance has not until recently been crucial to the understanding of the organization of work systems. The fact that social orders reflect and manifest an order of mind cannot be overstated.

Part of what is interesting about this research is that different theorists may arrive at different conclusions. The most simplistic conclusions involve a view of mind as a data processing mechanism that gathers quanta of data into progressively more inclusive sets of data. This view of mind is challenged when one replaces the concept of "data" with that of "information" inasmuch as information is a selection occurring within a field of data. That is the moment of realization, in a sense, that the organizational phenomena of Jaques' research has uncovered an ethical rather than a mechanical universe.

The vision of mind as data-organizer rather than information selector carries over in subtle ways toward various prescriptions on how to do things such as design a firm, a market, an academic discipline or a system of governance. Where one ends up theoretically appears to have much to do with what one is happy with. Is it good enough, for instance, that conflict is avoided, and that the social environment reflects peace and quiet? Maybe, but some prefer more rambunctious environments. At some point, it is a matter of variance in taste, and the kind of results one desires from a given design. To accommodate such variance, it is necessary to refrain from overstated design requirements. One might—for the same reasons—wish to avoid the distopias tending to accompany all attempts to override or simplify human diversity.

It seems that the more deeply we understand the human organism, the greater the likelihood we will successfully avoid its confinement within limiting orthodoxies. I therefore see no fault in theoretical attempts to map the mind, and to examine its boundaries, so long as such attempts are self-aware, and resist entrenchments. My use of the term "theology" is less of an attempt to establish a theology than to recognize the human propensity to make theories into theologies. One possible antidote is to devise a theology that tends to allow for the redefinition of boundaries. This has always been my view of what real science seeks to do.

Along these lines, I have left this book with a modest sense of conflict over what a spiritual theology demands of its promoter. Perhaps unsurprisingly, on the one hand I have supported an aggressive stance toward political constituencies intent on the enslavement of others, even though the same may be the only plausible way to establish peaceful social relations. On the other hand, I hold freedom out as a matter of choice to its adherents—whether as a theory or a theology—and advocate against the use of force as a means to its implementation. These are just what they appear to be—i.e., taken in the abstract, they are inconsistent positions.

This inconsistency is not accidental, and is meant to place emphasis on the fact that freedom frequently tasks us to decide what we love most. Some things are worth great risks and sacrifices, and admitting this is the case is an important part of what we mean in using "freedom" to denote something of value. Something within us causes us to establish

On Freedom: Postscript

it, and in its establishment, to resist its dissolution. This, at times, requires a sense of discrimination between what we will allow within our house, and what we will allow in the houses of our neighbors. It isn't always easy to tell the difference between what lies within and without, but telling the difference is part of the challenge, and not to be avoided merely because it is difficult.

Peter Gibson Friesen

Appendix:
The system of notation that marks the contours of a Logic of Maturation

Peter Gibson Friesen

I. The meaning of signs

The beginning of a logical calculus of development (aka a logic of "work") is, in a manner of speaking, a calculus that must in some sense explain beginning.

It is difficult to talk coherently about development without identifying a point from which development occurs. It doesn't need to be an absolute beginning, and the explosion of a singularity. Nor is it an atomic construct, where complex structures are built like figures from a set of exchangeable pieces. It is, rather, a calculus that allows us to examine the place from which building occurs, as one refinement after another is added to it.

A calculus of this nature would not be available but for the efforts of individuals who have taken an interest in the nature of thinking, and have developed languages to describe it—some of which are entirely symbolic. Here we adopt and examine the symbolic expression of beginning—not "a" beginning or "the" beginning—but beginning, in a calculus developed by G. Spencer Brown, in his Laws of Form. Since it may be regarded as a calculus of any number of different beginnings, it may alternatively be regarded as a calculus of "creation."

This calculus functions by calling attention to the fact that any given construct (object) comes to experience by way of selection from a "void." A void is not the same thing as "nothing." It refers to a substrate within which nothing has been selected, and is thus entirely neutral and valueless. The blankness of a piece of paper is not nothing, because it is indeed something to a marking device. It is a void that can hold the mark. The void the paper represents consists in its blankness, not in its nothingness.

We can call attention to the void either by leaving a space blank on this page, or perhaps to be clearer, to set aside a space within this page that is different than this page that is both blank and that has infinite extension—representing both its removal from this text and its extension beyond this page—with a dotted line.

This will do. But this includes the notion that a void is an imaginary postulation.

A void may hold any number of things, but since none are selected, it is, within any logic the mind can imagine, content-less. This is because content cannot exist without selection.

The effect of selection is a mark which sets the space represented in a void state apart from a space that has been selected. It can be any kind of boundary imaginable, so long as it cleaves as separate an object apart from void. The selected space can, within a void represented by a blank piece of paper, be a mark that holds that space separate and apart from the void. That mark works best—for a number of reasons—as a circle. Thus the circle marks a distinction.

This will do. 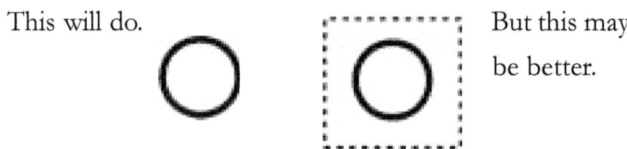 But this may be better.

They "mean" the same thing. The mark would mean nothing if the space within the mark were not in at least one respect different than the space outside. It may help to reinforce the difference by filling the space inside the mark, in order to emphasize that the inside has content, and the outside, by definition does not.

This can be rather confusing, because the content within cannot be seen as different than its boundary. This is not a trivial problem, and is the basis of a logical paradox known as Russell's paradox. But for now our objective is to communicate that in the distinction of an object this sets interest of distinction apart from its content. But this indicates the same concept.

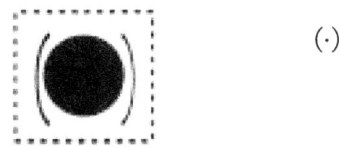 (·)

They mean the same thing, so long as one imagines that "( )" is the mark that sets a boundary around content "·" that is separated from a void. It is, moreover, important to realize that the space within a circle that is itself undifferentiated, is a singularity, because it is a single "place," and not an aggregation of places. The space inside the mark on a piece of paper has an infinite number of points within it, but this isn't just any space. The mark as beginning is meant to separate one space from another, and since there is no basis for stating difference within that space, the space itself can be represented as a singularity, a point. However, it is also meant to be representative of any distinction, and so it could be a complex or differentiated space, and we might therefore prefer to substitute "·" with a variable " Δ " to emphasize that it can be anything—an academic discipline or an eyelash. So this can be either:

 or (Δ)

There is within the symbolic language itself no meaningful difference between the use of pictures and the use of parenthetical symbols to encase a variable—where ( ) is used to indicate predicate to a term represented by a variable, as in Pm(Tn). This is another way of representing functions within a predicate calculus, and is commonplace among the available logical notations for that. So, what is different about what Brown is trying to represent with pictures? How does this logic stand apart from more traditional symbolic systems? Well, there are several differences.

II. The difference between this notation and conventional logic

1. First, symbol (depiction) on the left indicates something of a point of origin—here an object being the projective effect of an intervention on a void. It is, in other words, representative of an intersection between two axes, one of which standing alone is naught but a neutral substrate—a void, and the other is a projective connection to it, depicted as:

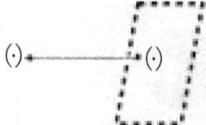

The difference is from this angle significant, in that it identifies a relationship between observer—an interested singularity (point of reference) and object (a projected singularity bound within interest). Brown thus places the observer in the logic, and thus presents a logic of creation. Again, it is not about any particular beginning, or the one and only beginning, but about beginning—i.e., it begins when the observer (subject) enters a void. It introduces a dimension to the void that was not there before, and that opens up spaces within it. It might also be depicted as follows:

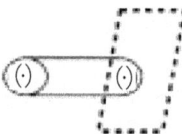

The logic thus presents an endogenous hierarchy—subject/interest/object, interest representing a relationship between subject and object, an interest having content. The hierarchy is endogenous, because the "direction" of hierarchy moves inward—from object projected on a void, to the observer's interest, to the observer. It may be educed from the object itself, where its content, radius and center are diverse aspects of a complex creation. This is a reason why a circle works well as a metaphor of containment, as opposed to, say, a rectangle.

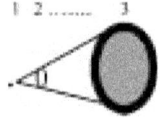

It is, perhaps, meaningful to note that "2"—representing "interest" remains active as governing principle throughout the course of assembling a construct or ordered field contained within a result. It is, as such, a productive endeavor, or process description, involving effort and motivation not unlike the scientific method, where hypotheses (states of interest) are tested. One might therefore say that the logic of creation is the same (or transfers easily to) a logic of work, investigation or science. Embedded within it is the notion of an extended present or "moment" that allows one to record the passage of time.

2. A second difference between this and conventional symbolism identified above is the transfer of limitation—evident in the spatial limitation of an enclosed mark to the parenthetical symbol. Thus the "( )" enclosing "·" acts as more than a separator of predicate from term, but is the predicate. The effect is a separator that has experiential content. Thus the distinction between predicate and term identified as P(T) is revealed as a redundancy, and can in fact be misleading if it fails to attribute "predication" to the sign ( ) itself.

This shift in the use of symbols has significant consequences, all of which pertain to the fact that the presence of interest (state of motivation or engagement) introduces enclosure to the space being demarcated from a void. It dictates what is allowed in the way of selection, and thus of distinction. To use Brown's terminology, the mark represents an "injunction"—i.e., a way of telling the user of the calculus what is allowed. He fits this concept within a generalized canon: "What is not allowed is forbidden."

The logic of creation, which is in one sense abstracted to "all" objects, is consciously and ironically self-limited by emphasizing that objects occur in a space constructed within a principle of containment. With this he derives two illustrations of the operation of this restriction.

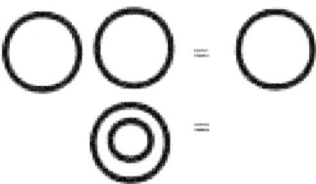

In the first instance we see that the multiplication of spaces is not allowed, and in the second, that the subdivision of spaces is not allowed. There is, in other words, no space to hold the aggregation of spaces, and a space selected cannot occur within a space selected without deselecting the original space. These are difficult concepts to master, because in the world of simple objects and spaces, we aggregate and subdivide with ease, but in a logic of development we attempt to imagine an observer engaged with an environment where the formation of a distinction requires all its attention, will and effort.

Another way to understand these laws is to imagine the difference between a substrate having only two possibilities, and another having three. In the first, the introduction of a distinction marks one state from another. But the repetition of singular distinction has no place to hold any but the distinction allowed. So it reduces to a single distinction. Similarly, the application of a distinction against that distinction has no place to hold it, and so it cancels the content of the first—nothing more. In a substrate allowing either of two states, in addition to an unmarked state, repetition can manifest in two places, and so is allowed. Distinction against the distinction can do one of either an unmarked state canceling the original, or in a second marked state standing beside the first. Negation within a plural substrate is, by definition, ambiguous. "Not light," can within a plural continuum, signify "dark," or "dim."

One can't, in other words, identify beginning without restricting what is allowed. Otherwise the end and the beginning can't be differentiated, and it would therefore not be a logic of creation. Conversely, one can't identify the course of development occurring through the expansion of what is allowed without being conscious of restriction at the moment of creation.

3. A third difference between this calculus and others preceding it is the recognition that the duality information theorists associate with binary codes does not inhere in the void (the substrate in which the observer selects), but is the consequence of a singular assertion within it. The existence of a binary code is entirely dependent on a monistic (one and only one) intervention occurring as a state of engagement, motivation or interest, which is coded as a sequenced transformation of an unmarked state to a marked state—from "0" to "1". The notions represented by "0" and "1" don't exist by themselves, rather "1" exists and establishes duality in a transformative process. That process may also be referred to as a "negation"—i.e., "negation" and "distinction" are equivalents.

What this means is that all of what we think of as "objective" reality, can be coded within a binary structure once we allow the aggregation and subdivision of spaces, applying "negation" as an operator of change—and accordingly, time. A computational machine can, in a manner of speaking, hold complex structures evident in an "objective" reality, but these structures are meaningless without an effective interface that allows an observer to select within it.

Additionally, all logics derived from a binary version of objectivity can proliferate within the logic of creation, including what we refer to as Boolean algebra—which other mathematicians have demonstrated is derivable from a binary origin. These derivations are qualified, however, by the fact that transformation is a conscious act functioning within constraints evident in the range of one's state of interest. A logic reflecting any actual state of consciousness or mind will need to tie a binary expression to its functionality. It is otherwise specious to think of the logic emerging from it as in any sense prescriptive—i.e., able to distinguish logical from illogical thinking.

4. A fourth difference is the acknowledgment of an iterative relationship between the observer and the product of observation. This is the point at which the formality of this logic both interfaces with and in a sense separates from knowledge being acquired in the behavioral sciences. This deserves some explanation.

This can be depicted through the modification of an earlier diagram, where the arrow of projection is regarded as bi-directional.

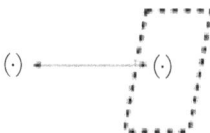

The diagram is a statement to the effect that what the observer selects produces changes in the void that affect the observer. This is a process of adaptation deriving from the perceived resistance of the void to selection. An interesting question is whether this resistance is in some sense definable within this calculus, or whether the process of adaptation is in that and other respects "extra-logical." The issue cannot be addressed without at least attempting to apply the calculus to it.

It is helpful along these lines to recognize that resistance may be redefined in terms of whether the void agrees or disagrees with the interest preceding the construction of a space. Agreement and disagreement are not, however, properly included in creation, but are something learned after selection takes place. One must learn of a resistant void, at which point one becomes more than an observer, and one's interest is about more than mere selection, but of selection that agrees with what one wants. This would be the basis of the judgments of "true" and "false," or rather assessments of whether the distinction agrees or disagrees with the interest brought to the void. Within a world of "true" and "false" the observer and the void undergo transformation. The observer becomes a "self" (seeker) and the void becomes an "environment" (resistant world). "true" and "false" are not part of creation, but are added to it. They are not logical constructs, but are better regarded as a subclass of behavioral science.

One might object that the existence of a formal structure cannot come into being of its own accord, because if indeed selection is activated by interest one must as an organism want something. Since desires develop over millions of years, beginning is more of an emergent phenomenon, and there is no coherent sense of beginning that in some sense precedes what an organism seeks to acquire from an environment. They may be right about that, but the abandonment of formalism does not disappear with a shrug of that nature.

An unconvincing way to deal with this objection is to attempt to redefine interest in terms of negations, negations of negations, etc., as it has for many centuries tended to sequester philosophy within esoteric languages, or worse, to pit philosophy against science. There are, nonetheless, several philosophically informative responses to the objection.

For one, there is a difference between the experience of desire and one's acknowledgment and acquiescence to it. The discovery that the world offers resistance is quite a significant event in the successful navigation of it. This is especially true of the social world, and the complex patterns of resistance and cooperation of a social surround. Ask any adolescent trying to get a date, and the frantic adjustment to priorities it entails. While the emergence of self out of the observer is driven by needs that are much more assertive than conceptual selection, the management of these needs does, in a manner of speaking, validate Brown's formalism, where distinction in all contexts begins with engagement.

On an entirely different level, given the fact that we are born with drives produced by genetic adaptation, we are tasked to consider whether that fact fits within or without Brown's formalism. Here it is useful to acknowledge that observer is not the same as a person, but ambiguously refers to any observational platform. On the scale of evolutionary biology, an observer may be regarded as a trans-generational relationship between a species and environment, where environment includes all that is given, including biologically determined states of want. The observer continues to stand apart from that, and makes its selections.

In more practically situated behavioral disciplines, where the development of a given individual faced with environmental challenges is being examined, one may come to appreciate a developmental process beginning with infancy and proceeding throughout childhood and adulthood. With each phase of development, beginning consists of a void that presents itself without selective intervention, which may include objects that were once the subject material of arduously constructed spaces that have become part of what is—without trying, seeking, or selecting. The question remains: how does one begin in dealing with it?

It is hard to say that Brown's formalism falters as a calculus of beginning, unless one develops a crabbed view of what a distinction and a void are. Part of the value of it is that the forms he developed tend to educate one's view of all processes that are developmental, and serve as a point of clarification for much of the behavioral sciences.

5. A fifth basis of distinction between Brown's formalism and others is that it is useful in identifying the limits of reductive thinking. As noted above, what we refer to as a "void" and "observer" occupy distinct dimensional axes, but they co-evolve. The problem with his calculus is that of the construction of a developmental logic without violating his canon: "What is not allowed is forbidden." Consider the three following diagrams.

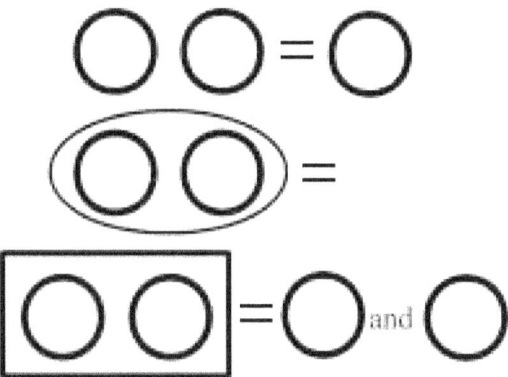

The first of these correctly states his principal of addition, essentially prohibiting the multiplication of spaces where the distinction is limited by its own design as precluding a space in which a distinction involving the same interest can aggregate. The second illustrates the reduction occurring when we reduce (·)(·) to (·) and then subject it to cancellation ((·))=. The third is designed to prompt a question of a formal nature. What if aggregation is allowed? This question is expressed as a hypothetical, what if we allow it with a secondary distinction focused on aggregation?

It is not a trivial question, merely because it asks the obvious. How does the observer break the enclosure that forces a distinction to gather into a single space—the simplification of a complex construct into either a void or a singularity (first and second options above)? The simple answer seems to be that one must coordinate, or rather, integrate and combine the assertion of one type of distinction with another. That would mean, however, that the observer carries (in the third instance) two types of interest to the void. One of this is "primary"—i.e., it marks the interface of observer in a state of engagement with the void. The other is "secondary"—i.e., it marks the interest of the observer in remembering the products of that engagement. Without the second, the gathering signified by the connective "and" won't happen.

Within the logic of creation, aggregation is prohibited without consciously adding to what is allowed. The effect ought to be clarity brought to bear on the creative effect of a given distinction, and thus its limitation. It is not clear, however, that Brown desires, or believes in this limitation.

III. Does Brown's calculus violate his canon?

One way to avoid the limitation revealed in Brown's logic of creation is to introduce the concept of indeterminate value, by replacing the ambiguity over value cast into a void with a variable sign "a", "b" or "c"...., such that indeterminacy prevents simplification. This changes the diagrammatic landscape considerably. This is, in fact what Brown suggests.

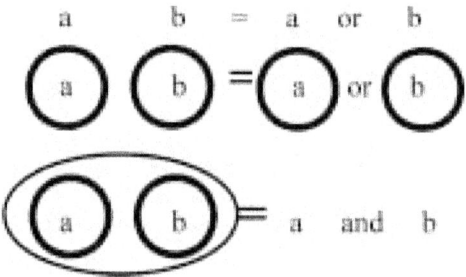

This is the Boolean reduction—i.e., that the conjunctive expression "a and b" is the same as to say "not(not a or not b)." Apparently, this is an elegant method of transition from a concept of distinction that prohibits aggregation to one that allows it. The support seems to rest in the redefinition of a "void" from one in which one and only one selection can be made to one in which many can be made, so long as the many are viewed as potentialities—potentiality being represented by an undetermined variable. It is not at all clear however that this distinction can be made (void as substrate of singular distinction and void as a holder of multiple indeterminate distinctions) without the addition of an interest—an addition that is only implicitly allowed by populating the void with variants.

It seems that the elegance of this transition occurs not through the application of the originating distinction "( )", but through the application of a secondary interest in a void consisting of an aggregate of potential selections. The view of a void as a holder of potentialities (variables) is allowed because of an interest in aggregating them. Thus, the more direct way of expressing this is to add a secondary interest in aggregation to the original distinction—indicated in the preceding page as a symbolic translation or definition "[( ) ( )] = ( ) and ( )". The reduction displayed above tends to conceal the introduction of this as a second interest, inasmuch as "variable" is unaccompanied by explanation that it implies a secondary distinction. Brown has allowed a second type of distinction without formally announcing it.

In this respect, his formalism is ambiguous in its reduction because it attempts to define an expansion in terms of a singular type of distinction—here to bring about existence through distinction (negation), and to collapse distinctions of a different type into the original. This way of looking at creation leads one into an esoteric universe. For example, one might define "self" as a distinction of a distinction, "other" as a distinction from self (distinction of distinction of distinction), and thus proceed on into topologically infinite domains formed out of negations. These are of potential interest to a few, but in the end are philosophically unsatisfying. The solution to this problem, it seems, is to admit to the diversification of logical interests—and tether them to "methods" or "modes" of logical processing.

One might say that Brown never intended to use a singular "distinction" as the basis for the formation of a universe, but rather that he uses distinction as a mental activity that allows for the formation of a universe. Logic is not generative, but permissive inasmuch as behavioral phenomena occur within it. It is, perhaps, a legitimate interpretation, but not a preferred interpretation, given that the laws he identifies with creation are redolent with existential value. It appears that where Brown succeeded—and succeeded magnificently—was in identifying a common thread in the commencement of engagement with any neutral substrate imaginable—of creation. He did not identify a process that extended meaningfully beyond the act of creation, except to suggest that all

On Freedom: Appendix

extensions are acts of creation, and thus occur as motivated interventions. This addition allows us to better appreciate what he set out to do.

This is an observation, not a criticism, and is meant in large part to avoid the misinterpretation of his calculus. To some extent, the problem we are looking at has a partial solution. The solution consists in the ambiguity of the magnitude of a distinction. A distinction framed within this calculus works equally well as a simple singularity or as a compound singularity.

Consider the following diagram, where "Δ" represents a quantum of extension.

⊙

The extension could be an infinitesimal (simple and unvariegated), as we noted above, in which case the difference between the "interest" that marks the boundary and the content of the interior cannot be depicted, except to use a symbolic representation where boundary stands apart from the singularity of its content—i.e., (·). But "·" can also identify a complex and variegated content, that extends or radiates from the observer as "measure." The difference between the two might be depicted with symbols expressed as follows:

$$(\cdot) \longrightarrow \{O\}$$

The second instance is allowable if and only if the interest represented by { } is of a different kind than the interest evident in ( ). We have allowed, in other words, a universe to exist on another order of magnitude than the first, and for an act of creation to occur as an expanded sense of creation evolved from the first. In the second, it is a populated void, each with complex content embraced (selected) by interest. One way to imagine the second image is in terms of a boundary of variation within which complex objects can form. It is one thing, for example, to construct "a" face, and quite another to engage a face within a range of acceptable content. That range is the "form" within which complex parts are consolidated. The boundary indicated by "()" is the same as the boundary indicated by "O" except in the latter instance the boundary is content engaged by a new interest, where in the first, boundary is a state of engagement. In the latter instance, boundary can be examined as an object fused with points organized within it.

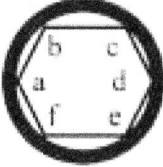

There are a number of interesting extrapolations to draw from this interpretation of Brown's logic of creation. One is that creation not only allows expansion, but encourages it, and places it before mind as an event of discovery, and that the events associated with movement from one event of creation to another are not only aggregative, but consolidative—i.e., that mind not only gathers its creations, but organizes them into larger objects. Consolidation terminates at the exact moment of a new creation. It repeats itself along a continuum. There is within this calculus some need to identify and sequence the steps that lead from one order to another, though to be fair to Brown, there are some suggestions on how to go about it.

IV. Getting from concatenation to organization

We have come to the realization that allowing concatenation (the aggregation of distinctions) occurs when a second distinction is allowed that adds to the one identified with the moment of creation. A distinction of a different "type" thus represents the addition of a way of grouping distinctions—a square about circles or [(·)(·)]—which was implicit in the change of the void from a single region from which one distinction can be made to a void in which many distinctions can be gathered together. At this stage there is no meaningful difference between a void offering itself as a gathering of points or places and a void offering itself as a gathering of variables—as locations or contents not yet discerned, but presumably different from each other. Their difference consists in the fact that as they are taken from the void, they are subtracted from it. So, assuming a void consists of the following points,

……… or abcdefgh

their spatial (or other) relationship to each is not a matter of consideration. They do not exist in a configuration (shape) or even a line (sequence).

The distinction of one "·" or "a" simply subtracts that point from the void, and what remains of it is "......."(seven points) or "bcdefgh". The process of concatenation is only a matter of securing selections as a matter of recollection such that [...(·)] leaves ...., or [abc(d)] leaves efgh. The point is: the differentiation of distinctions as variables is allowed only because we have introduced a second "type" of interest that manifests as a second logical process acting upon a void whose character has changed because the interest intervening upon it has changed. Doing more than this is not yet allowed—i.e., an assessment of how one point links to another, or of how a group of points is organized is not allowed unless and until an interest is added that brings a new type of distinction into the navigational horizon.

So, how do we make this transition? It appears to start with an interest in whether one point includes, by its nature, another. This interest is quite different than inclusion in a space by way of concatenation—i.e., by way of bringing one point together in memory or recollection. Rather, it involves interest in the way one point "·" in some sense "belongs with," "implies," or "intimates" another. To this purpose, we have options. One might attempt to represent this with a new distinction "<>" to represent inclusion endogenous to the points themselves.

$$[(<\cdot\cdot>)] \text{ leading to } [<\cdot\cdot><\cdot\cdot>(<\cdot\cdot>)]\cdot\cdot$$

It is a coherent construct, and identifies the addition of a "type" of distinction, but it also introduces a matter of confusion stemming from a failure to annotate the difference between the point distinguished in the first instance "(·)" from the point it implies. For this reason, and perhaps others, the sign (·) requires an indication of separation from the point it implies. Thus, as perhaps a better alternative, we can depict the distinction as:

$$<(\cdot)\cdot> \text{ or } (\cdot)<>\cdot \text{ or } (\cdot)>\cdot$$

Important to this depiction is that "· >·" does not equate with the expression "(·)·" or "(a)b" also known as a conditional expression "if a then b", because the conditional expression does not, of itself, identify one and only one point that follows after it. That would be an inappropriate construct within Brown's system of notation, because in order to identify one and only one "b" implied from "a" one would have to introduce a new distinction that limited the manifold of variables following after "a". After selecting "a" within a void consisting of abcdefgh..., what follows thereafter is any one of bcdefgh...—or rather, (a)bcdefgh. One cannot, in other words, derive implication from negation ( ) and concatenation [ ] without allowing for the introduction of a new interest that links one point to a single other.

Despite the fact that we are looking at a void that now includes the concept of linkage between points, we are not in a position to organize the points into what might be described as a structured whole. What we have is a potential series extending from any given point, and a void thus conceived as a gathering of implicative strings. So, from any given starting point, any number of implicative strings might emerge.

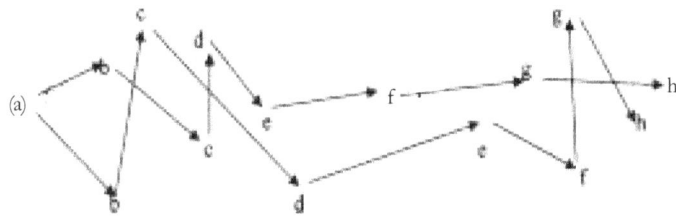

These strings are, in a real sense, formless constructs, yet they could not emerge without the introduction of a distinction that allows points to be linked. The void, once conceived as a blank slate from which a mark could be drawn, has since changed to a concatenation of potential variants, and now as a tangle of linked series that can be draw from any given point in the void. The void has become a tangled fabric, a place where strings emerge from dust—a cobweb. It can only be so because we have allowed it to be, in and through the introduction of a new kind of distinction.

This point of arrival reveals—to a degree—how changes in interest accompany changes in the way a "void" is conceived. It was once pretty simple, only to become like a jungle of vines. However, without a perception of disorder, one can hardly become interested in order where strings get compacted into wholes—complex singularities. Order, in other words, is not an arbitrary formulation, but emerges from a full appreciation for a void bound within its own ligatures. Order is in a strong sense discovered, but only after the complexity of a world of possibilities is fully appreciated—a world that in some sense resists organization, but also allows it.

It should be appreciated that the "seeing" of strings is a matter of recollecting links that have been drawn between points—of aggregating or collecting a number of "heres" and "theres" where "there" becomes a platform of extension—another "here" to another "there". The mind so engaged is very careful to keep "here" and "there" distinct, because one cannot effectively engage "there" without knowing or supposing "here." Such is the way that mind can reach into and distinguish places within a void without knowing them, or rather, by doing so imaginatively. But this does not occur without first having taken interest in imaginative extension in the first instance, and of conceiving a void now as a substrate that is linked to and extending from a preselected point of observation.

This appreciation allows (but does not require) mind to engage in the next distinction that brings "organization" into a void that had previously been conceived as a concatenation of linked points. This occurs as an act of mind that dislodges the observer from a given point of observation. The observer, in other words, reflects upon its point of observation, and is thus able to conceptualize a void as an aggregation of points of observation. In so doing, the observer is in a position to reform its point of observation from a single point to a complex of points—i.e., from a simple singularity to a complex singularity viewed as a structured whole.

One way to represent this is to take a distinction identified as a "string" of implications, and introduce a new type of distinction that identifies ordered complexes among the aggregated strings.

A distinction of this nature cannot occur simply by bringing points together, but (1) by bringing them together in a way that by knowing the point of observation of one, all are known and (2) by bringing them together without destroying or disrupting the links making up the strands following from each point. To enact a change of this nature, it is necessary to remove or extract a gathering of points of observation into a single point of observation conceptualized as a compound construct—an ordered whole. In the stranded diagram above, such a construct is not allowed because any given point of observation links to something regarded as exterior to it. It is, in that sense, "linear" because there is but one direction toward which attention is cast—from "here" to "there." But let us assume that thought is at some new point of comprehension, interested in the difference between two different points of extension, both linked to a first point of observation.

Peter Gibson Friesen

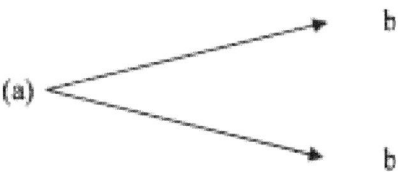

It is apparent at this time that the case of "b" above is not the same as the case of "b" below, but also that both cases of "b" link to "(a)" simultaneously. In order to separate one case of "b" from another they are assigned different "names"—accomplished by deciding to call one "b" and another "c", each occupying now a different position in a "space" which holds them together in such a way that their separate positions can be recognized. This change is radical, inasmuch as we have changed "c" from something that happens after "b" to something that happens along with "b." The variables "b" and "c" have thus been moved from outside a's space to its inside. The variable "a" thus shares its space with other points of observation, and the relative positions of "b" and "c" are known from "a", just as the positions of "a" and "c" are known from "b."

This transformation may be viewed as a process or consolidation or "organization" depicted through a modified sense of what lies within the void. So, what we have is a point of observation "(a)" aware that it is linked simultaneously to every point in the void and thus formulating ordered constructs within it. It is a process of deconstruction and reconstruction. Looking at this:

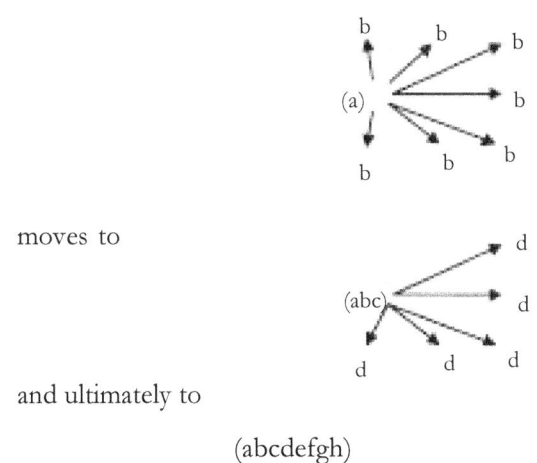

moves to

and ultimately to

(abcdefgh)

where each point is located within a single space in which each point occupies one and only one place, and achieves location within that space because of its relationship to the other points within it. The position of "a" is a matter of being in relation to others in a state of co-occupancy of that space.

We have, unfortunately left this concept of organization without a mark of identification, and, having simply assembled them into the same space, failed to symbolically identify the kind of assembly it is. So, take the symbol ><, to set this distinction apart from implication <>. The distinction >< indicates organization or order expressed as a point of observation formed as a compound singularity—or rather (→abcdefgh←).

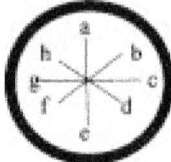

Interestingly, the consolidation of a void into an ordered complex occurs within a domain defined by a state of motivation—which when projected onto paper becomes the value of a distinction. The process of consolidation takes this state as a "limit" within which a consolidated whole grows.

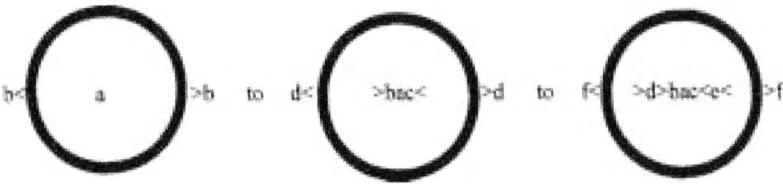

And so on.

This might be a way of expressing the emergence of hierarchy—i.e., as the progressive expansion of ordered boundaries within a functional value. This understanding of hierarchical expansion strongly suggests a moment of "impact," so to speak, where the content of the motivating value acts as a constraint on growth, where it had up to a point served as an activating principle on growth.

V. Going back to singularity

The challenge of allowing the addition and integration of processes that extract meaningful combinations of spaces from a void begins with creation, and thus evolves much as Brown suggests—i.e., by allowing them to occur, formally acknowledging what has been allowed, and examining the effect of such allowance. In this instance they have led us toward the realization that the same thing which activates intervention into field also defines the limits of the field itself. A point within the field becomes reflective, where each space is defined in terms of the principle of order applicable to the others. Individuality is defined as "aspect" of a whole, or it is otherwise not defined.

That one's engagement with the void occurs through a process of reflection converting disorder into order, and chaos into navigable wholes, is an important feature of our use of symbols to represent logical growth. That is to say, the selection of a whole out of a void has much to do with the fact that we are individualized beings, and that ordered cohesion is something projected into space, and likewise reflects back into what we are. This, it appears, is one of Brown's most important contributions to logic, inasmuch as the logic formally acknowledges the individuation of the observer as integrated with its logical creations. Indeed, without a motivated observer, nothing forms at all.

This tendency noticeably widens a pathway between logic and cognitive science, inasmuch as it acknowledges that learning and logic are interconnected, that learning does not occur instantaneously or without struggle, and that it is imbued with ethical considerations. Moreover, we can, by examining an exchange, of sorts, between the formation of objects and the formation of identity, develop working hypotheses in conflict with the view that the human mind branches outward capriciously. Its consolidative nature is not only part of growth, but in some sense enables it.

The completed consolidation of a given void thus becomes a single space—not a concatenation of spaces, not a strand, and not an ordered hierarchy, but the type of thing that can be embraced by a value. Take the picture we drew before.

$$(\cdot) \longrightarrow \{O\}$$

From this image, we can by process of recollection, acquire a fuller appreciation of a mutual dependency between what is often referred to as "mind and body"—of a sense of "body" that changes with the intervention of mind, and of mind that is in a real sense bound and reactive to its creations.

Endnotes

¹ While spatial metaphors are not perfect, they are in many instances superior to the use of words. Here, for example, is depicted a distinction between "mind" as a subject matter of investigation and "mind" as a presence that performs the investigation. One is object and the other subject. How might one otherwise clarify that the word "mind" can mean two different things, and likewise encourage one to exercise caution in treating them as one and the same? Particularly in an essay devoted to the examination of something like "freedom" a distinction such as this, marked by the spatially intensive symbol, ( ), is a device by which one separates a subjective universe that, among other things, forms and delimits objects, not the least of that is a "thing" it refers to as "mind." It is relatively easy to acknowledge that there will always be a nexus established between a given theory of the object "mind" and the theories devised to explain a natural world. What is more difficult to comprehend, and thus what justifies the use of visual metaphors, is that change in the subject "mind" is apt to change one's view of matter. Scientific freedom challenges one to imagine exactly what sort of change of subjective orientation might produce a phenomenon that is now referred to as "science."

² A good summary of the findings of this research, along with a rich analysis of its applicability to the biological and behavioral sciences occurs in The Life and Behavior of Living Organisms, Elliott Jaques, Praeger Publishers, 2002.

³ The change of words here serves in part to make the comparison between information processing and the assertion of value more obvious. It also displays something of the way scientific vocabularies progress.

On Freedom: Footnotes

[4] This movement was founded in the second half of the nineteenth century by Mary Baker Eddy as the Church of Christ Scientist. She founded a publishing society that prints a daily newspaper with a good reputation for accurate and unbiased reporting.. This society also prints monthly journals, that constitute an impressive archive of testimonials to healing, most of which are authenticated by sworn statements of witnesses. She authored a textbook entitled <u>Science and Health</u> offered to the public.

[5] One ought to take Wittgenstein's admonition seriously, and not speak of things that cannot be spoken. Not that he was an irreligious person, in that he privately confessed reverence toward matters of spirituality, and for that reason supported caution in the use of words to describe an infinite and transcendent divinity. This attitude is consistent with scientific freedom, that opposes the human tendency to make God into something that resembles many of the more primitively fashioned human thoughts and values. This book will not attempt to theoretically prove the existence of God, nor will it attempt to prove that extraordinary events often attributed to divine intervention occur. Testimony to such events, however, encourages scientific and philosophical thought to view self-validating logical and moral systems skeptically, or to reshape them in such a way that they cannot validate themselves, and thus indicate respect toward whatever spiritual unknowns may be thought to influence one's life.

[6] Propositional Logic has been carefully organized and arranged to make this hierarchical progression accessible, and that progression has been placed beside research into the work behaviors in stratified organizations. See R.O. Gibson an D.J. Isaac, "Truth Tables as a Formal Device in the Analysis of Human Actions" in <u>Levels of Abstraction in Logic and Human Action</u>, Elliott Jaques Ed., Heineman Educational Books, Ltd., London, 1978. This essay on logic benefited considerably from their work, and is founded on it. The first essay converting this logic to a notational language that described information search strategies was completed by this author in 1979 and these are later referred to as "information processing" by Elliott Jaques at some point in the latter half of the 1980s.

[7] The tendency to extrapolate from a given theory of cognition toward a scale of "intelligence" has in modern times produced a number of

distortions that adversely affect the capacity of society to choose and educate leaders. This may be a consequence of limited resources, and of a need to preserve the appearance of impartiality in personnel selection, but tends to minimize the importance of personally motivated inquiry while maximizing conformity to standardized "tests" of intellectual proficiency. Overall, there appears to be disrespect for the lability of intellect as a creative response to interest, as well as a premature institutional identification with incompletely conceived measures of aptitude. An emphasis on intellectual freedom thus supports caution in the "measurement" of intelligence and places more value on the development of curiosity, because intellectual function serves interest.

[8] One might prefer to express this presence with a circle "O" and argue that presence reflected in capacity has "extension" and thus use that symbol as a place of origin for a logic. While there is some elegance to that idea, there is also a potential for confusion, inasmuch as a circle conveys more than extension, but also self-awareness associated with the separation of an inside and outside. A circle assumes a formative history consisting of the assembly of many places. It knows it is a compound presence even though it asserts itself as a singularity into the void.

If one is, in reading the paragraph above slightly bewildered, then the argument of this paragraph is in a sense proven, i.e., the fact that much more can be said of a circle than a point suggests that while it shares the symmetry of a point, it introduces structure that deserves special attention. We will soon examine the phased development by which points—here representing simple objects—are assembled into circles—here representing compound or complex objects.

[9] The use of the term "distinction" and an accompanying symbol "O" to establish a logically primary idea borrows from G. Spencer Brown's Laws of Form. It is a great work. Among other things, it is a logically complete calculus, which develops from a very simple foundational idea, i.e., "distinction" (also referred to as a "cross" or "mark") versus "void" (the "unmarked" state). This appears to be another way of juxtaposing existence and nonexistence.

Value is thus a prerequisite to existence +, which is incapable of standing apart from the blankness of nonexistence holding a magnitude of 1, except to satisfy an expectancy introduced by a conscious agency. The presence of interest ( ), however, is not necessary to suppose the absence of existence. Absence can occupy a void whether interest is present or not. Since it is nothing, it does not depend on an expectancy to be nothing. Interest does not make something out of nothing, but may ascertain something + where it may have otherwise been supposed nothing.

[10] Since an object is a complex construct over which one labors, it has magnitude reflecting the capacity and interest of the person putting it together. Many objects require little effort, and are accordingly taken for granted. A tree, for example, is an object distinguished from the manifold of objects with little effort, though at one point in life, perhaps at a certain stage of infancy, to see a tree as an object separate from others requires much effort, and evokes sentiments of fascination. As one enters into the world which demarcates employment, as a hunter, farmer, mechanic or craftsman, the objects with which one is preoccupied may be things like skills, environments, and the composition of materials.

[11] While the criticism of dualistic systems of logic is sound, the term "Aristotelian" is placed in quotes out of respect to scholars of Aristotle who take issue with the assertion that Aristotle's logic—which is dispersed throughout a number of his works—is as his followers restated it from time to time. There does appear to be a legitimate criticism of a logic which treats the division of the world into classes as final as opposed to provisional terminology. Such fallacies are more difficult to find in the works of great philosophers than many suggest, though it does appear that Aristotle frequently exhibited overconfidence in the power of the intellect to understand the world through the generation of hierarchic terminologies in science. (See Korzybski, <u>Science and Sanity</u>, Fifth Ed. 1973)

[12] "Truth Tables as a Formal Device in the Analysis of Human Actions," R.O. Gibson and D.J. Isaac, <u>Levels of Abstraction in Thought and Human Action</u>, Eliot Jaques, Ed. Heineman Press, 1978, at p. 159.

[13] The technical definition of "reflexivity" is that a given relation R is reflexive on a given set of objects A, if for each object "a," aRa. A relation of order is not reflexive on a set unless the relation establishes one and only one position for each member. If a reflexive relation were

turned into a command, it would function as a rule of exclusion, i.e., a given object may exist in a symmetric or anti-symmetric relation, but not both. Thus the expression (>···<·>·) is prohibited, and should thus be expressed as (>···<)·>·. An exception to a rule of order does not belong in a relation of order. Understanding understands all that fit into one pattern, and those that do not fit are not understood.

Reflexivity stated as a rule reinforces the formation of hierarchically organized systems. When one states, for example, "you are not a member of my division, but you are a member of my organization," he establishes a hierarchy. The expanded view ruptures the reflexive characteristic of the smaller view. One's comprehension of the relationship of a system's members undergoes expanding revision until the "whole" is "understood." The same is true between a leaf, a branch, a limb, and a tree, or between a family, a community, a society, and a nation. It is difficult to organize ones' perceptions of nature without hierarchical groupings, the hierarchies one imposes work better for some natural phenomena than for others.

However the hierarchy that represents a theory or pattern of organization is neither true nor false, and is evaluated in terms of its usefulness in satisfying a given interest. Understanding, or rather, theoretical behavior, seeks only to comprehend a related system of objects because such comprehension tasks its entire capacity. One is reaching out toward the boundaries of his field of labor, and is thus apt to express irritation upon interruption. There is urgency toward completion, while the task of evaluation is for others.

[14] One must be cautious while attempting to maintain focus on a notion such as "self," "ego," or "identity." The tendency to conceptualize an interest or expectancy as something like an object led into previously mentioned referential paradoxes. The challenge is significantly greater when one deals with identity which, like "gravity" is "imagined" through its effects, or if taken for granted, is often known for the sense of fragmentation that accompanies its loss. Therefore the acknowledgment that identity is a different "type" of experience than interest, is in a sense foundational to the comprehension of morality and moral freedom. If one admits that a person's "interests" tend to gather around something, then that something is what he supposes to be their identity.

[15] The use of "property" as the means or mechanism of maintaining security against enemies would tend to highlight the command "you shall

not steal." A problem with this extension arises from the simple fact that Moses identified this as the third and not the second of the five moral commands. The second reads "you shall not commit adultery." One could say that Moses did not intend that these come in any particular order, but that tends to ignore the fact that these commands have been recited with such frequency over thousands of years that there would be a tendency to organize them on an intuitive scale. Another explanation might be that in the culture of Moses, as well as in the tribal cultures that followed, family organized under the covenant of marriage has been regarded as the basis of security, and that property had an entirely different meaning than it presently has. Property could in a different culture include a sense of interdependency, and responsibilities for sharing and distribution that implicate relationships of trust more significantly than marriage. If something of this nature is true, one may consider whether the movement from a tribal to a mercantile system of production involves a major change of where one locates, in moral terms, the assertion of a given interest. It would mark a significant point of social transformation.

So, it might make more sense to view the moral commands of Moses as arranged in order of importance, with the consolidation of family as "closer" to life than the one's accumulated possessions. Such considerations might move the order of rules from the developmental sequence we have been interested in.

One might also prefer to refrain from reading in to such commands matters of depth or spiritual resonance, and treat them as one would behavioral "rules of thumb." This approach is neither satisfying nor pragmatic, because it tends to ignore the mythic propensity of consciousness to redefine terms, or otherwise to view terms metaphorically. Morally loaded terms such as "steal" or "adulterate" are ascribed meaning for the sake of convenience, and in an important way, mean what one needs for them to mean in order bring a moral value to a point where it can be experienced. This, as mentioned earlier, tends to distinguish mythic formations from intellectual formations, i.e., myths as instrumentalities manipulated to evoke moral sensation. In the modern world, it seems that "steal" more aptly responds to one's interest in security and that "adultery" deals with intimacy, but this writer must admit that he cannot say exactly what they meant to Moses.

[16] In the chapter that addressed Intellectual freedom, emphasis rested on a "calculated risk" associated with the mental act by which one includes another object without knowing it, and that the sizing of that object as

one equal in magnitude to the one that is known was inseparable from the "validity" of the extension. In the case of Moral freedom, the sense of risk that attends to one's embrace of other can hardly be overstated. Particularly where one has managed to find self-satisfaction in a world he defines in terms of "good" and "evil" the prospect of having evil appear in one's own "house" is quite unsettling. That "good" becomes less of a comfort to him and the "evil" is felt as a cancerous organ that cannot be excised without killing its carrier. The resolution comes as "good" and "evil" are recognized as vanities, and as that which was once infested with evil is not more or less than the good one makes of it. To one acting with genuine altruism, there are unsettling questions, such as "am I a fool?," "am I a bleeding heart?," "will I be appreciated?," and "will it be taken in the spirit it is given?" These are resolved as one decides, "it doesn't really matter." He has, in a sense, crossed over that boundary in the course of defining his own good in a way that accommodates the impassivity and/or resistance of others.

[17] This is an adaptation from a model conceived by Elliott Jaques, in his book The Form of Time, Crane Russak Heineman, 1982, which makes a distinction between time as a series of events and time as a span of comprehension, the latter of which is manifest in the presence of consciousness in a material world. The reality one attempts to imagine without mind has no places in it because there is no presence in which "place" satisfies an interest. Thus it has occurred to this author that it may be productive to assess what foundation in consciousness there is for a continuum of places, and how that might have derived from a structured or segmented consciousness in that the occupancy of a "place" includes an awareness of other places.

[18] I have taken this name applying to a third "level" of language from a paper written by my professor of philosophy—Nathaniel Lawrence—which he submitted to the International Society of Time in 1978. He used the term to apply to a level of language where the unifying element of human experience, or "identity," is melded to the language event.

[19] This observation is that of the legal philosopher, Ronald Dworkin (Deceased 2013) in his famous argument commonly referred to as "the right answer thesis."

Author Biography

Peter Friesen received his Bachelor of Arts from Williams College in 1978, majoring in Philosophy. He continued to pursue his philosophical interests in Public Administration at the University of Southern California, and a Juris Doctorate at the University of California, Hastings, awarded in 1982. While a student at the University of Southern California, he studied with Elliott Jaques, and developed the first logical descriptors for cognitive development utilizing an information processing model. Following his J.D., he continued to work under a grant from the Mead Foundation in 1983, producing a manuscript identifying symmetries between levels of logic, morality, language and politics. This manuscript leaned heavily on the work of R.O. Gibson in the formulation of "levels" in first order logic (who was then a collaborator with Jaques) and notational system formulated by G. Spencer Brown in his ground breaking work, Laws of Form.

At the same time, Friesen developed a law practice focused at first on criminal trial work, and then moved to civil trials in cases against corporate and governmental institutions. His casework included lead roles in employment cases, police misconduct, defamation and unfair competition; he was recognized with four San Diego Outstanding Trial Lawyer awards and named on the list of America's Best Lawyers.

In 2000, he took three years away from his law practice to finish the work he started with Jaques in 1979, producing a manuscript entitled On Freedom. As a result of the manuscript, Friesen was invited to serve on the Board of Advisors of Requisite Organization International, Inc. (ROII) He is currently an advisor to Global Organization International, founded by Ken Shepard to serve a continuing international clientele developing and applying organizational science. He has since been working on the completion of a work suitable for publication, various applications of the theory of organization it represents, and has continued to practice law on matters related to employment, business organization and the management of policy.

He has also completed a manuscript on the application of organizational science to the field of Economics, entitled <u>Law and Economic Order: A Theory of Requisite Economy</u>. Friesen sees his role as one that can assist in the development of productive communication between Logical, Psychological, Legal and Economic disciplines, and as part of this role, to assist in the marginalization of influences that are interested in disrupting this communication. The book <u>On Freedom: Organizational Science Examined Philosophically</u> is the first of Friesen's works for which he has sought publication.

www.ingramcontent.com/pod-product-compliance
Lightning Source LLC
Chambersburg PA
CBHW051121160426
43195CB00014B/2292